Solar Energy: Systems, Technology and Engineering

Solar Energy: Systems, Technology and Engineering

Darren Erickson

www.statesacademicpress.com

States Academic Press,
109 South 5th Street,
Brooklyn, NY 11249, USA

Visit us on the World Wide Web at:
www.statesacademicpress.com

© States Academic Press, 2022

This book contains information obtained from authentic and highly regarded sources. All chapters are published with permission under the Creative Commons Attribution Share Alike License or equivalent. A wide variety of references are listed. Permissions and sources are indicated; for detailed attributions, please refer to the permissions page. Reasonable efforts have been made to publish reliable data and information, but the authors, editors and publisher cannot assume any responsibility for the validity of all materials or the consequences of their use.

ISBN: 978-1-63989-489-5 (Hardback)

Trademark Notice: Registered trademark of products or corporate names are used only for explanation and identification without intent to infringe.

Cataloging-in-Publication Data

Solar energy : systems, technology and engineering / Darren Erickson.
   p. cm.
Includes bibliographical references and index.
ISBN 978-1-63989-489-5
1. Solar energy. 2. Renewable energy sources. 3. Engineering.
I. Erickson, Darren.
TJ810 .S65 2022
621.47--dc23

Table of Contents

Preface   VII

Chapter 1   Sun and Solar Energy   1
- Sun as an Energy Source   1
- Solar Radiation   12
- Solar Energy   43
- Solar Energy Storage System   57

Chapter 2   Solar Collector and Thermal Technologies   67
- Solar Collector   67
- Solar Pond   99

Chapter 3   Principles of Solar Energy Generation   111
- Energy from Solar Spectrum   111
- Solar Thermal Power Generation   115

Chapter 4   Solar Power and Photovoltaics   128
- Solar Power   128
- Photovoltaics   145
- Photovoltaic System   163
- Concentrated Solar Power   185
- Concentrator Photovoltaics   195

Chapter 5   Solar Energy Devices   200
- Solar Cooker   200
- Solar Street Light   207
- Solar Water Heating   210
- Solar Thermal Pump   221
- Solar Refrigeration   223
- Solar Distillation   225
- Solar Dryer   228

Permissions

Index

Preface

The radiant heat and light received from the sun is known as solar energy. It is a renewable source of energy that is classified in to active solar and passive solar depending upon how the energy is captured and distributed. Passive solar technologies are those which don't use any external mechanical or electrical device and depends upon natural circulation of air. Whereas, active solar systems use devices such as fans and pumps to circulate air or fluid through solar collectors. Some prominent examples of solar technologies are photovoltaics, solar architecture, solar heating and artificial photosynthesis. Solar energy has undergone rapid developments in the last few decades and is expected to enhance sustainability, reduce air pollution, lower the fuel prices and mitigate the effects of global warming in the coming years. This book attempts to understand the multiple branches that fall under the discipline of solar energy engineering and how such concepts have practical applications. Most of the topics introduced in it cover new techniques and the applications of solar energy engineering. This book will serve as a valuable source of reference for those interested in this field.

A foreword of all chapters of the book is provided below:

Chapter 1 - The radiant light and heat produced by the sun makes it a potential source of energy if properly harnessed. The solar electromagnetic radiation emitted by the sun is called solar radiation which can be harnessed and stored for future use. This chapter discusses solar energy and the numerous storage systems in a comprehensive manner; **Chapter 2** - Any device that is used to collect solar radiations is called a solar collector. There are several different types of solar collectors available for use, each with its own working principle. The technology of producing, storing, controlling, transmitting and getting work done by heat energy is known as thermal technology. Both these topics are discussed in detail in this chapter; **Chapter 3** - In order for the solar radiations to be used, it has to be captured, stored and then converted to electricity or any other form of power for consumption. The solar radiation can be harnessed in two different ways. Solar light can be used in power cooking appliances, lamps, panels or batteries. Solar heat can be used for heat generation and powering solar thermal stations to produce electricity. This chapter provides a detailed study of solar energy generation and its principles; **Chapter 4** - Simply put, solar power is the use of the sun's energy either directly as thermal energy or its conversion to electricity through the use of photovoltaic cells. The conversion of the sun's energy into electricity is known as photovoltaics. The set-up used to undertake this process of electricity generation is called a photovoltaic system. This chapter has been carefully written to provide the reader with a better understanding of the subject matter; **Chapter 5** - The availability of modern solar appliances has resulted in the diverse applications of solar energy. From household devices to industrial sized machineries, solar energy can be used in diverse ways. Solar cookers, solar street lights, solar water heating systems, solar thermal pumps, solar refrigeration, solar distillation, solar dryer, etc. are some applications of solar energy which have been thoroughly discussed in this chapter.

At the end, I would like to thank all the people associated with this book devoting their precious time and providing their valuable contributions to this book. I would also like to express my gratitude to my fellow colleagues who encouraged me throughout the process.

Darren Erickson

Chapter 1

Sun and Solar Energy

The radiant light and heat produced by the sun makes it a potential source of energy if properly harnessed. The solar electromagnetic radiation emitted by the sun is called solar radiation which can be harnessed and stored for future use. This chapter discusses solar energy and the numerous storage systems in a comprehensive manner.

Sun as an Energy Source

The sun is a star close to earth that provides the energy to our planet. It is the ultimate source of all the energy sources. The Sun is the largest object in our solar system. The radius of the sun is 6.9×10^8 m that is 109 times that of the earth. The mass of sun is 2×10^{30} kg which is around 300,000 Earths.

Anatomy of a Star

The anatomy of a star consists of:

- Chromosphere,
- Photosphere,
- Convection zone,
- Radiation zone,
- Core.

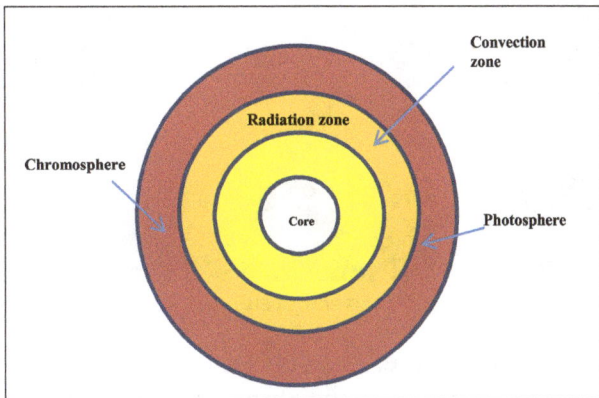

Anatomy of a star.

The energy emission in the sun is by nuclear fusion. The centre or core of the star plays the platform for fusion process. After the energy production, it is released away from the core by means of radiation in the radiation zone of the star.

- Radiative Zone: The radiation zone extends out to about 0.7 solar radii from the core. In this zone the radiation makes it hot and it helps in transfer of heat from core in outward direction. The ionic forms of hydrogen and helium emitting photons in this zone travel a short distance before being reabsorbed by other ions. The temperature of approximately 7 million kelvin at closer to the core is reduced to 2 million at the boundary with the convective zone. There is also reduction in density with a range of 20 g/cm³ closest to the core to 0.2 g/cm³ at the upper boundary.

- Convective Zone: This zone has lower temperature than in the radiative zone and non-ionized heavier atoms. This zone lies in 200,000 km from the surface. The rising thermal cells carry the majority of the heat outward to the Sun's photosphere. Once these cells rise to just below the photospheric surface, their material cools, causing their density increases. This forces them to sink to the base of the convection zone again – where they pick up more heat and the convective cycle continues.

As the energy moves away from the core and passes through the radiation zone, it reaches the part of the star where the energy continues its journey towards the surface of the star as heat associated with thermal gradients. This part of the star is called the convection zone. The surface of the star, called the photosphere of thickness ~500 km, emits light in the visible part of the electromagnetic spectrum. The star is engulfed in a stellar atmosphere called the chromospheres of thickness ~2500 km. The chromosphere is a layer of hot gases surrounding the photosphere.

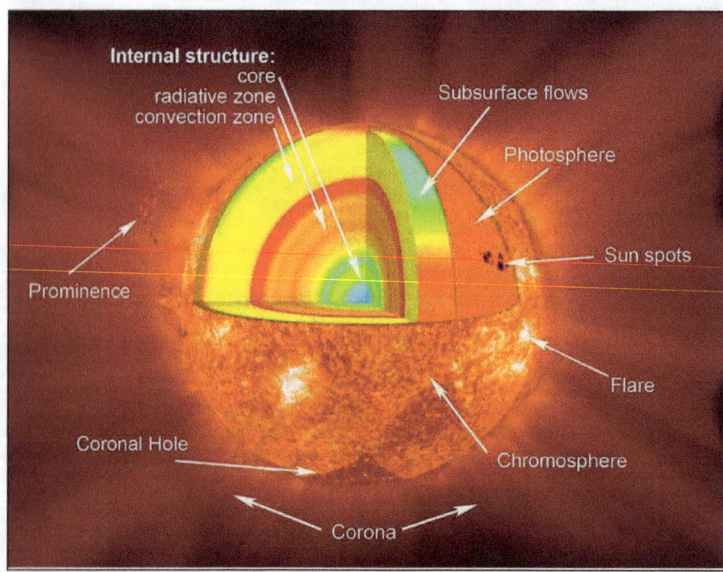

Interior of sun.

The Corona is the Sun's outer atmosphere. The sun's magnetic fields rise through the convection zone and erupt through the photosphere into the chromosphere and corona. The eruptions lead to solar activity, which includes such phenomena as sunspots, flares, prominences, and coronal mass ejections.

Atmosphere of Sun

The atmosphere of the sun is composed of several layers, mainly the photosphere, the chromosphere and the corona. It's in these outer layers that the sun's energy, which has bubbled up from the sun's interior layers, is detected as sunlight.

The lowest layer of the sun's atmosphere is the photosphere. It is about 300 miles (500 kilometers) thick. This layer is where the sun's energy is released as light. Because of the distance from the sun to Earth, light reaches our planet in about eight minutes.

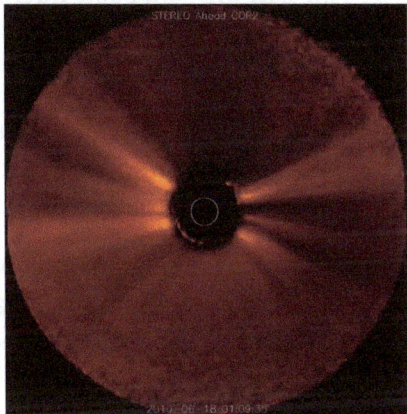

Image of the solar corona.

The photosphere is marked by bright, bubbling granules of plasma and darker, cooler sunspots, which emerge when the sun's magnetic field breaks through the surface. Sunspots appear to move across the sun's disk. Observing this motion led astronomers to realize that the sun rotates on its axis. Since the sun is a ball of gas with no solid form, different regions rotate at different rates. The sun's equatorial regions rotate in about 24 days, while the polar regions take more than 30 days to make a complete rotation.

The photosphere is also the source of solar flares: tongues of fire that extend hundreds of thousands of miles above the sun's surface. Solar flares produce bursts of X-rays, ultraviolet radiation, electromagnetic radiation and radio waves.

The surface Gas Pressure (top of photosphere) is 0.868 mb and Pressure at bottom of photosphere (optical depth = 1) is 125 mb. The effective temperature is 5772 K. Temperature at top of photosphere is 4400 K and at bottom of photosphere is 6600 K. Temperature at top of chromospheres is ~30,000 K.

Composition of Photosphere

Major Elements	
H	90.965%
He	8.889%
Minor Elements	
O	774 ppm
C	330 ppm

Ne	112 ppm
N	102 ppm
Fe	43 ppm
Mg	35 ppm
Si	32 ppm
S	15 ppm

The next layer is the chromosphere. The chromosphere emits a reddish glow as super-heated hydrogen burns off. But the red rim can only be seen during a total solar eclipse. At other times, light from the chromosphere is usually too weak to be seen against the brighter photosphere.

The chromosphere may play a role in conducting heat from the interior of the sun to its outermost layer, the corona. "We see certain kinds of solar seismic waves channeling upwards into the lower atmosphere, called the chromosphere, and from there, into the corona."

The third layer of the sun's atmosphere is the corona. It can only be seen during a total solar eclipse as well. It appears as white streamers or plumes of ionized gas that flow outward into space. Temperatures in the sun's corona can get as high as 3.5 million degrees Fahrenheit (2 million degrees Celsius). As the gases cool, they become the solar wind.

Core of the Sun

The core of the Sun occupies 20–25% of the solar radius from the centre. The energy production by fusion of hydrogen atoms (H) into molecules of helium (He) occur at the core. The high energy production is the result of extreme pressure and temperature that exists within the core. The pressure and temperature of the core of the sun is equivalent of 250 billion atmospheres (25.33 trillion KPa) and 15.7 million kelvin, respectively. The 99 percent of energy produced by sun happens in 24 percent of the radius i.e. core. The heat energy is transferred to other parts and outer space from the core.

The core is the source of all the Sun's energy. Fortunately for life on earth, the Sun's energy output is just about constant so we do not see much change in its brightness or the heat it gives off. The Sun's core has a very high temperature and the material in the core is very tightly packed or dense. It is a combination of these two properties that creates an environment just right for nuclear reactions to occur.

The intense heat at the core of a star prevents an electron from remaining bound to any one atomic nucleus to form atoms as we know them. Instead, the negatively charged electrons and positively charged nuclei move randomly about independent of one another. This kind of gas is known as a plasma. There are equal amounts of positive and negative charge, so the plasma itself has no net charge. The majority of nuclei in the Sun are protons, more familiar to us as nuclei of hydrogen atoms. Next most common is the helium nucleus composed of two protons and two neutrons, bound together by a force far stronger than any electric field: the nuclear binding force. While electric fields are not strong enough to keep atoms together at the high temperature in the core, the nuclear binding force is easily strong enough to keep every helium nucleus bound tightly together.

As the positively charged nuclei of a plasma move randomly around, pairs will often approach close together, but electric repulsion keeps them from actually touching. At higher temperatures the nuclei move faster and will approach each other more closely. The temperature at the core of the Sun is high enough that pairs of protons can occasionally approach close enough for the nuclear binding force to attract them together, overwhelming their electric repulsion. After this occurs once, a few more encounters will result in four protons combining into a single helium nucleus. (During this process, different nuclear effects convert two of the protons into neutrons and several more exotic sub-atomic particles, as shown in the figure). This is known as nuclear fusion, and it releases a net energy due to the tight binding of its constituents. The continuous fusion of protons into helium nuclei releases the energy that powers the Sun.

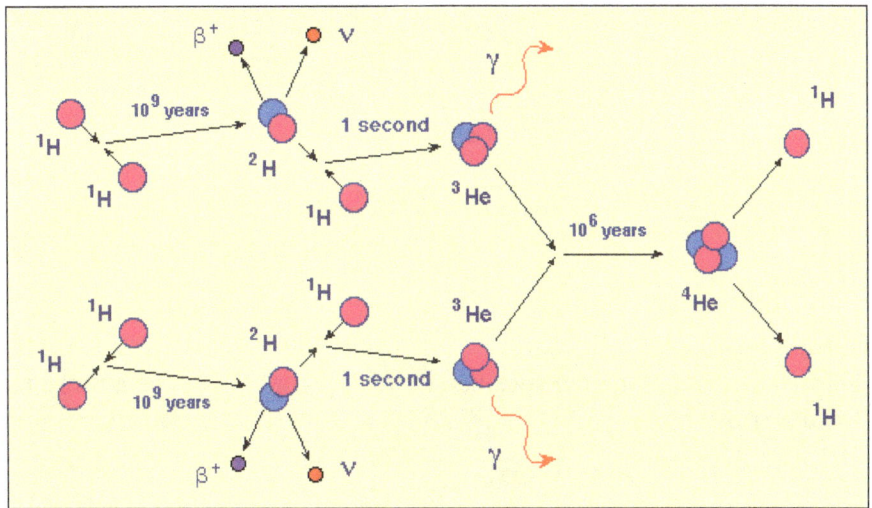

Nuclear Fusion Reaction in the Sun

The temperature inside the sun is > 10,000,000 K. The stellar nuclear reactions are responsible for energy production inside the sun. The hydrogen atoms combine together to form helium atom by the nuclear fusion reaction. The Sun releases energy at a mass–energy conversion rate of 4.26 million metric tons per second, which produces the equivalent of 38,460 septillion watts (3.846×1026 W) per second.

These reactions are of two types:

- Proton-Proton Chain (P-P chain),
- CNO cycle.

The Sun emits 4 x 10^{26} Watts of power.

Proton-Proton Chain Reaction

It converts Hydrgen atoms (H) to Helium (He). This stellar reaction is most efficient in lower mass stars like the Sun. The reaction results in fusion of four protons into one alpha particle with the release of two positrons and two neutrinos (which further changes two of the protons into neutrons) and energy.

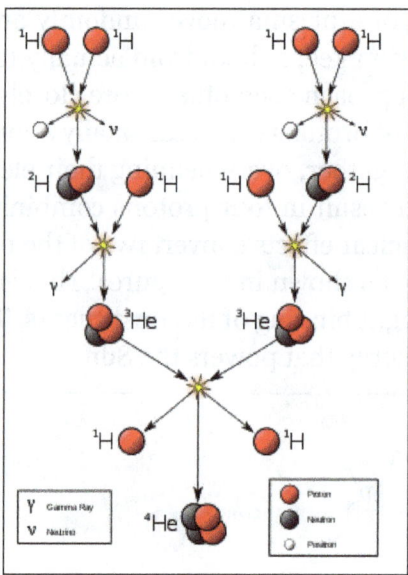

Proton-proton chain reaction.

CNO Cycle

This reaction is important in heavier stars. In this cycle, hydrogen is used to synthesize heavier elements. The heaviest elements are synthesized by fusion that occurs as a more massive star undergoes a violent supernova at the end of its life, a process known as supernova nucleosynthesis.

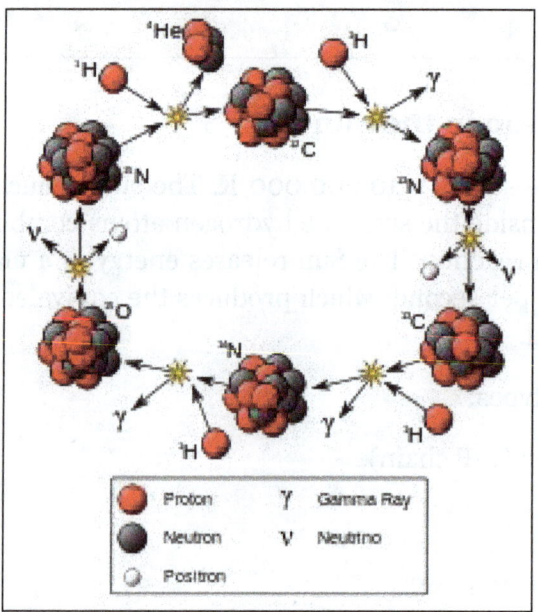

CNO cycle.

This is most efficient in heavier stars having temperature greater than 16,000,000 K. This cycle was proposed by Hans Bethe in 1939.

The sun remains hot due to equilibrium maintained between Nuclear reactions; chemical or gravitational energy. But it is inadequate to maintain luminosity for billions of years. Pressure balance (gas pressure vs. gravity) and energy balance (production vs. outflow) are both needed to regulate heat energy inside the sun.

Energy Production in Sun

The energy released during conversion of hydrogen to helium is mainly in the form of photons. The energy released can be estimated from the mass difference.

Mass of $4H^1$ = 6.69048 x 10^{-27} kg

Mass of $1He^4$ = 6.64648 x 10^{-27} kg

Δm, mass difference = (1) - (2) = 0.04400 x 10^{-27} kg

$E = mc^2$ = 4.0 x 10^{-12} Joules per reaction

If 10% of hydrogen is converted into helium for the sun, it will produce 10^{44} J of energy which is sufficient to keep sun burning for 10 billion years.

Calculations Related to Energy Production in Sun

Examples:

1. How much mass is converted to energy every second in the sun to supply energy to the earth?

Solar constant = 1372 W/m²

R_{earth} = 6.4 x 10^6 m

$D_{earth-sun}$ = 1.5 x 10^{11} m

M_{sun} = 2.0 x 10^{30} kg

To get solar constant in units of mass (kg), we have,

$$\frac{1372 \text{ J/s}}{C^2} = \frac{1.372 \times 10 \text{ kg m}^2/\text{s}^2}{9 \times 10\ 16 \text{ m}^2/\text{s}^2} = 1.524 \times 10^{14} \text{ kg/s}$$

The total energy falling on the earth is solar constant times the projected area of the earth (from the sun), then,

A earth SC = (1.3 x 10^{14} m²) x (1.5 x 10^{-14} kg/s/m²) = 2 kg/s

Which is the mass converted every second to supply light incident on earth.

2. How much metric tons of H are converted to He to supply this energy?

Take,

m_p = 1.67262 x 10^{-27} kg

$D_{earth-sun}$ = 1.5 x 10^{11} m

M_{He} = 6.64648 x 10^{-27} kg

The mass difference between four protons (6.69048 x 10^{-27} kg) and one helium nucleus (2 proton +2 neutrons) is 0.04400 x 10^{-27} kg.

The ratio of H-burned to converted mass is 6.69048/0.04400 = 150 or 0.7 % of rest energy (mass) of original H converted to He.

Hence total mass burn rate of the sun is,

$$150 \times 4.2 \times 10^9 \text{ kg/s} = 6.3 \times 10^{11} \text{ kg/s} = 630 \text{ million metric tons each second.}$$

3. How long will the sun continue to produce energy? (Ignoring other processes and emissions from the sun.)

Dividing the burn rate into fuel supply (total mass) gives an estimate of how long sun will last.

Life time of sun = M_{sun} = 2.0 x 10^{30} kg = 3.2 x 10^{18} s = 10^{11} yr.

Burn rate = 6.3 x 10^{11} kg/s

The sun convert only 10 % of its H mass to He, so the above estimate o = is high by a factor of 10 and main life time of sun is ~ 10^{10} yr.

Energy Transport in Sun

The energy produced at core of the sun by fusion reaction is transported by Conduction, Convection and Radiation.

Table: Energy transport in sun.

Part of sun	Temperature (10^6 K)	Density (g/cm^3)	Energy transport
Core	~15	100	Convective
Radiation zone	~3	1	Radiative
Convective zone	~1	0.1	Convective

The stars generally use the convective and radiative method.

Solar Energy to Earth

The sunlight that reaches the earth depends on the revolution of the earth around the sun.

Orbital plane: The orbit of the earth around the sun lies in a geometrical plane called the orbital plane. The ecliptic plane is the plane of the orbit that intersects the sun. The line of intersection between the orbital plane and the ecliptic plane is the line of nodes. Planetary orbits, including the earth's orbit, lie in the ecliptic plane.

The luminosity of a star is the total energy radiated per second by the star. The amount of radiation from the sun that reaches the earth's atmosphere is called the solar constant. The solar constant varies with time because the earth follows an elliptical orbit around the sun and the axis of rotation of the earth is inclined relative to the plane of the earth's orbit. Distances between points on the surface of the earth and the sun vary throughout the year.

The flux of solar radiation incident on a surface placed at the edge of the earth's atmosphere depends on the time of day and year, and the geographical location of the surface. Some incident solar radiation is reflected by the earth's atmosphere.

The fraction of solar radiation that is reflected back into space by the earth-atmosphere system is called the albedo. The solar radiation is absorbed or reflected by clouds (20%), atmospheric particles (10%), and reflection by the earth's surface (5%). Thereby prevents from reaching the earth surface.

The solar flux that enters the atmosphere is reduced by the albedo. So the solar radiation undergoes absorption, reflection or scattering by air, water vapor, dust particles, and aerosols while travelling to earth. The solar radiation that reaches directly to earth is direct solar radiation and that reaches after scattering is known as diffuse radiation.

Light is electromagnetic radiation. It can be used to transfer energy by the propagation of electromagnetic waves from one location to another. Two other energy transfer processes are conduction and convection. Conduction is the transfer of energy as the result of a temperature difference between substances that are in contact. Convection is the transfer of energy by the movement of a heated substance.

Solar Radiation Spectrum

The Sun's emission in the extreme ultraviolet part of the solar emission spectrum.

The Sun emits radiation from X-rays to radio waves, but the irradiance of solar radiation peaks in the visible wavelengths. Common units of irradiance are Joules per second per m^2 of surface that is illuminated per nm of wavelength (e.g., between 300 nm and 301 nm), or $W\ m^{-2}\ nm^{-1}$ for the plot below. These units are the units of spectral irradiance, which is also simply called irradiance, but as a function of wavelength. To get the total irradiance in units of $W\ m^{-2}$, the spectral irradiance should be integrated over all the wavelengths.

Note the following for the solar spectrum:

- About half of the energy is in the visible wavelengths below 0.7 μm. We can tell this by doing a quick integration.
- O_3 and O_2 absorb much of the UV irradiance below 300 nm high in the atmosphere.
- About 70% of the visible irradiance makes it all the way to sea level.
- O_3 absorbs a little of the visible irradiance.
- A significant fraction of the visible irradiance is scattered by clouds and aerosol. Some is reflected back out into space so that this portion never deposits energy in the Earth system.
- There are large wavelength bands in which water vapor, CO_2, and O_3 absorb infrared irradiance.

For solar wavelengths at which the absorptivity is high, the solar irradiance at sea level is small. Note that the big absorbers of infrared irradiance are water vapor, carbon dioxide, and ozone.

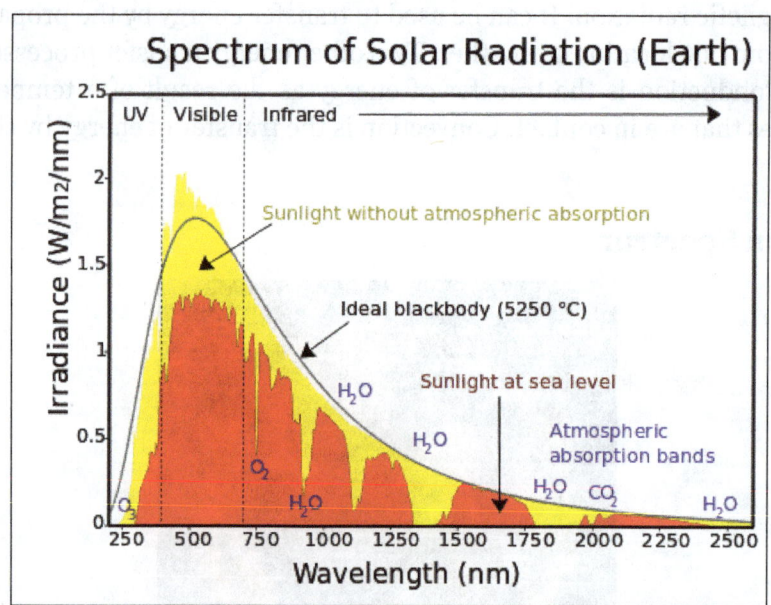

Solar spectrum and atmospheric absorbing gases from 240 nm to 2.5 μm wavelengths.

Sun Earth Relationship

The distance between sun and earth is measured in Astronomical Unit (AU). One AU is the distance travelled in 8.31 minutes at the speed of light. The mean distance of sun from Earth is 149.6 x 10^6 km with minimum of 147.1 x10^6 km and maximum of 152.1 x 10^6 km.

The earth receives almost all its energy from the Sun's radiation. Sun also has the most dominating influence on the changing climate of various locations on Earth at different times of the year. The Earth rotates about on a fixed plane that is tilted 23.5° with respect to its vertical axis around the sun. The Earth needs 23 hrs 56 mins to complete one true rotation, or one sidereal period, around the sun. The solar day, on the other hand, is the time needed for a point on earth pointing towards a particular point on the sun to complete one rotation and return to the same point. It is defined as the time taken for the sun to move from the zenith on one day to the zenith of the next day, or from noon today to noon tomorrow. The length of a solar day varies, and thus on the average is calculated to be 24 hrs. In the course of the year, a solar day may differ to as much as 15 mins. There are three reasons for this time difference.

- The earth's motion around the Sun is not perfect circle but is eccentric;
- The Sun's apparent motion is not parallel to the celestial equator;
- The precession of the Earth's axis.

For simplicity, we averaged out that the Earth will complete one rotation every 24 hrs (based on a solar day) and thus moves at a rate of 15° per hour (one full rotation is 360°). Because of this, the sun appears to move proportionately at a constant speed across the sky. The sun thus produces a daily solar arc, which is the apparent path of the sun's motion across the sky. At different latitudes, the sun will travel across the sky at different angles each day.

The rotation of the earth about its axis also causes the day and night phenomenon. The length of the day and night depends on the time of the year and the latitude of the location. For places in the northern hemisphere, the shortest solar day occurs around December 21 (winter solstice) and the longest solar day occurs around June 21 (summer solstice). In theory, during the time of the equinox, the length of the day should be equal to the length of the night.

The average time the earth takes to move around the sun in approximately 365 days. This path that the earth takes to revolve around the sun is called the elliptical path.

Equinoxes and Solstices

Equinoxes happen when the ecliptic (sun's apparent motion across the celestial sphere) and celestial equator intersect. When the sun is moving down from above the celestial equator, crosses it, then moves below it, that point of intersection between the two planes is when the Autumnal Equinox occurs. This usually happens around the 22nd of September. When the Sun moves up from below the celestial equator to above it, the point of intersection between the sun and the celestial equator is when Spring (Vernal) Equinox occurs. It usually happens around the 21st of March. During the equinoxes, all parts of the Earth experiences 12 hours of day and night and that is how equinox gets its name as equinox means equal night. At winter solstice (Dec), the North Pole is inclined directly away from the sun. 3 months later, the earth will reach the date point of the March equinox and that the sun's declination will be 0°. 3 months later, the earth will reach the date point of the summer solstice. At this point it will be at declination -23.5°. This cycle will carry on, creating the seasons that we experience on earth.

The earth is tilted 23.5°, so is the ecliptic, with respect to the celestial equator, therefore the Sun maximum angular distance from the celestial equator is 23.5°. At the summer solstice which occurs around 21st of June, the North Pole is pointing towards the sun at an angle of 23.5° as shown in figure. Therefore the apparent declination of the sun is positive 23.5° with respect to the celestial equator. At the Winter solstice which occurs around 21st December, the North Pole is pointing away from the sun at an angle of 23.5°. Therefore the apparent declination of the sun is negative 23.5° with respect to the celestial equator.

Change of Seasons

Seasons are caused by the Earth axis which is tilted by 23.5° with respect to the ecliptic and due to the fact that the axis is always pointed to the same direction. When the northern axis is pointing to the direction of the Sun, it will be winter in the southern hemisphere and summer in the northern hemisphere. Northern hemisphere will experience summer because the Sun's ray reached that part of the surface directly and more concentrated hence enabling that area to heat up more quickly. The southern hemisphere will receive the same amount of light ray at a more glancing angle, hence spreading out the light ray therefore is less concentrated and colder. The converse holds true when the Earth southern axis is pointing towards the Sun.

Sun Apparent Movement

From the heliocentric point of view, the Earth rotates and revolves around the sun in a counter clockwise direction. However, when we look at the Sun on earth, it appears to be moving in a clockwise direction. This phenomenon is known as the apparent motion of the sun.

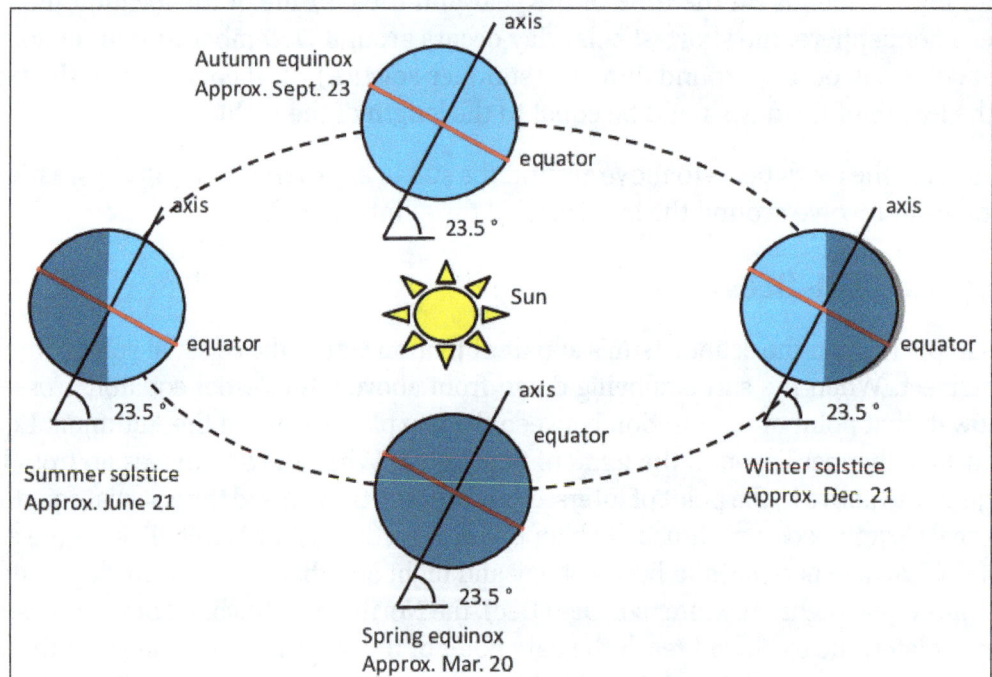

Solar Radiation

The energy in the sun is produced by nuclear fusion reaction and it maintains a high temperature at the surface. The radiated heat energy from the sun is called solar energy or solar radiation. This radiation is the fundamental part of all the biotic and abiotic processes on the earth. The energy and various gaseous cycles between terrestrial and atmospheric continuum depends on solar radiation. The water cycle, local weather and climate, circulation of wind are also controlled by the solar radiation. The photosynthesis is possible in the presence of solar radiation and hence it is important for plants and to animals.

The sun is considered as a black body having high surface temperature (5,800 °C). The electromagnetic spectrum of solar radiation ranged from gamma rays (100 nm) to radiowaves (1 mm). The solar radiation spectrum consists of emission at various wavelengths but more at short wavelengths.

The maximum emissive power of the radiation takes place at wavelength of 0.48 um. The average time taken by solar radiation to reach the surface of the earth from Sun is about 8 minutes.

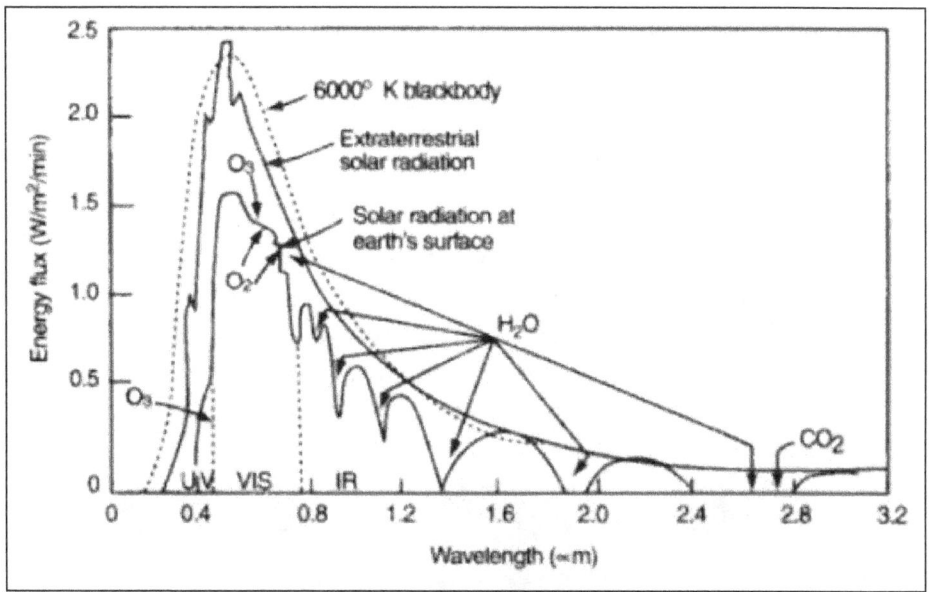

Solar radiation spectrum.

The solar radiation spectrum represents continuous spectra of electromagnetic waves. About 95 percent of the total energy lies in the short wave length region of 0.3-2.4 μm and about 99 percent is within 0.25-4 μm range. The long waves in the spectrum ranged greater than 4 μm accounts only to 1 percent. And the radiation spectrum from the earth consists mainly of long wavelengths and maximum emission takes place at 10 μm.

Irradiance and Irradiation

Irradiance is the rate at which radiant energy is incident on a unit surface area. It is the measure of power density of sun light falling per unit area and time. It is measured in watt per square meter. Heat energy is measured in joules and while watt or joules per second is unit of power.

Irradiation is solar energy per unit surface area which is striking a body over a specified time. Hence it is integration of solar illumination or irradiance over specific time (usually an hour or kilowatt a day). It is measured in kilowatt-hour or kilowatt day per square meter.

Irradiation = irradiance x time period

For example, if irradiance is 20 k W/m² for 5 h, irradiation is 20 × 5 = 100 k W/m².

The earth revolves around the sun in an elliptical orbit as shown in figure. The earth is closest to the sun on 21 March and 23 September. The earth is farthest from the sun on 21 June and 22 December. The mean distance of the earth from the sun is 1.495 × 10¹¹ m.

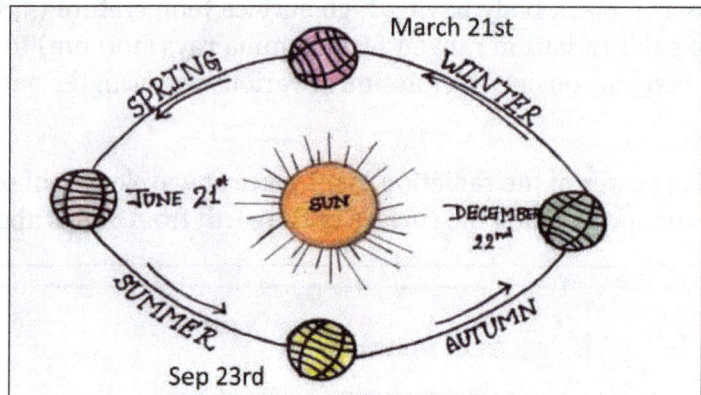

Elliptical orbit of earth around sun.

The intensity of solar radiation outside the earth's atmosphere reduces with distance and it is dependent on the distance between the earth and the sun. In fact, the intensity of solar radiation reaching outside the earths atmospheric varies with the square of the distance between centers of the earth and the sun. This is the reason why earth receives 7 % more radiation on 21 March and 23 September as compared to 21 June and 22 December. The intensity of solar radiation keeps on attenuating as earth propagates away from the surface of the sun, but the content of wave length in the radiation spectrum does not change.

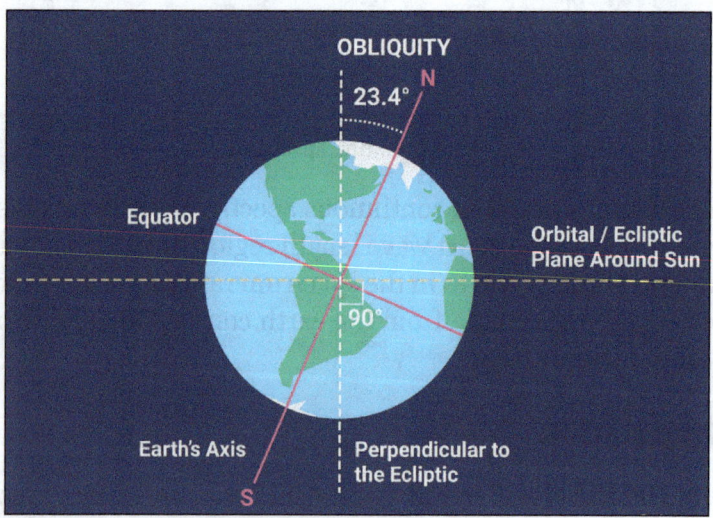

Inclination of earth's axis.

The earth axis is tilted about 23.45° with respect to earth's orbit around the sun as shown in figure owing to this tilting of earth's axis, the northern hemisphere of the earth points towards the sun in the month of June and it point away from the sun in the month of December. However, earth's axis remains perpendicular to the imaginary line drawn from the earth to sun during the month of September and March. The sun- earth's distance varies during earth's rotation around the sun, thereby varying the solar energy reaching its surface during revolution, which bring about seasonal changes. The northern hemisphere has summer in the month of September and March, both the hemisphere are at the same distance from the sun and receive equal sunshine. During the summer, the sun is higher in the sky, while the sun is lower in the sky during winter for the northern hemisphere.

Extraterrestrial Radiation

Solar radiation incident on the outer atmosphere of the earth is called extraterrestrial radiation. The extraterrestrial radiation varies based on the change in sun- earth's distance arising from earth's elliptical orbit of rotation. The extraterrestrial radiation is not affected by changes in atmospheric condition.

Almost the entire extraterrestrial radiation incident to the Earth has been radiated by its closest star, the sun. The rate of influx is called the solar constant, and it has a magnitude of 2 cal/cm^2/min. About half of the incoming energy occurs as relatively short "visible" radiation within a wavelength band of about 0.4–0.7 μm, and half is composed of "near-infrared" wavelengths between about 0.7 and 2.0 μm. There is a virtually perfect energetic balance between the quantity of electromagnetic energy that is incident to the Earth and the amount that is ultimately dissipated back to outer space.

Terrestrial Radiation

When radiation passes through earth's atmosphere, it is subjected to the mechanism of atmospheric absorption and scattering depending on atmospheric conditions. Earth's atmosphere contains various constituents, suspended dust and solid and liquid particles, such as air molecules, oxygen, nitrogen, carbon dioxide, carbon monoxide, ozone, water vapour and dust. Therefore, solar radiation or intensity of radiation is depleted during its passage through the atmosphere. The solar radiation that reaches earth's surface after passing through earth's atmosphere is called terrestrial radiation.

The major isotopes of concern for terrestrial radiation are potassium, uranium and the decay products of uranium, such as thorium, radium, and radon. Note that, terrestrial radiation includes an external exposure caused by these radionuclides.

These radionuclides are in trace amounts all around us. When the Earth was formed, a number of radioactive elements were formed. After the four billion years, all the shorter-lived isotopes have decayed. But some of these isotopes have very long half-lives, billions of years, and are still present. This radionuclide is known as primordial radionuclides and contributes to the annual dose to an individual. Because most of the natural radioactive isotopes are heavy, more than one disintegration is necessary before a stable atom is reached. This sequence of unstable atomic nuclei and their modes of decays, which lead to a stable nucleus, are known as the radioactive series.

All, natural radionuclides are usually divided into two groups depending upon their origin:

- Primordial radionuclides: Primordial radionuclides are radionuclides found on the Earth that have existed in their current form since before Earth was formed. Primordial radionuclides are residues from the Big Bang, from cosmogenic sources, and from ancient supernova explosions which occurred before the formation of the solar system. Bismuth, thorium, uranium and plutonium are primordial radionuclides because they have half-lives long enough to still be found on the Earth. Potassium-40 also belongs to primordial nuclides.

- Cosmogenic radionuclides: Cosmogenic radionuclides are those which are continually being produced by the interaction of cosmic rays.

Dose from Terrestrial Radiation

Low levels of uranium, thorium, and their decay products are found everywhere. Some of these materials are ingested with food and water, while others, such as radon, are inhaled. The dose from terrestrial sources also varies in different parts of the world. Locations with higher concentrations of uranium and thorium in their soil have higher dose levels. The average dose rate that originates from terrestrial nuclides (except radon exposure) is about 0.057 µGy/hr.

The major isotopes of concern for terrestrial radiation are uranium and the decay products of uranium, such as thorium, radium, and radon. Radon is usually the largest natural source of radiation contributing to the exposure of members of the public, sometimes accounting for half the total exposure from all sources. It is so important, that is usually treated separately. The average annual radiation dose to a person from radon and its decay products is about 2 mSv/year and it may vary over many orders of magnitude from place to place.

Radon: Health Effects

Radon is a colorless, odorless, tasteless noble gas, occurring naturally as the decay product of radium. All isotopes of radon are radioactive, but the two radon isotopes radon-222 and radon-220 are very important from radiation protection point of view:

- Radon-222: The radon-222 isotope is a natural decay product of the most stable uranium isotope (uranium-238), thus it is a member of uranium series.

- Radon-220: The radon-220 isotope, commonly referred to as thoron, is a natural decay product of the most stable thorium isotope (thorium-232), thus it is a member of thorium series.

It is important to note that radon is a noble gas, whereas all its decay products are metals. The main mechanism for the entry of radon into the atmosphere is diffusion through the soil. As a gas, radon diffuses through rocks and the soil. When radon disintegrates, the daughter metallic isotopes are ions that will be attached to other molecules like water and to aerosol particles in the air. Therefore all discussions of radon concentrations in the environment refer to radon-222. While the average rate of production of radon-220 (thoron) is about the same as that of radon-222, the amount of radon-220 in the environment is much less than that of radon-222 because of significantly shorter half-life (it has less time to diffuse) of radon-222 (55 seconds, versus 3.8 days respectively). Simply radon-220 has lower chance to escape from bedrock.

Radon-222

Radon-222 is a gas produced by the decay of radium-226. Both are a part of the natural uranium series. Since uranium is found in soil throughout the world in varying concentrations, also dose from gaseous radon is varying throughout the world. Radon-222 is the most important and most stable isotope of radon. It has a half-life of only 3.8 days, making radon one of the rarest elements since it decays away so quickly. An important source of natural radiation is radon gas, which seeps continuously from bedrock but can, because of its high density, accumulate in poorly ventilated houses. The fact radon is gas plays a crucial role in spreading of all its daughter nuclei. Simply radon is a transport medium from bedrock to atmosphere (or inside buildings) for its short-lived decay products (Pb-210 and Po-210), that possess much more health risks.

Radioactive Series in Nature

Radioactive series (known also as radioactive cascades) are three naturally occurring radioactive decay chains and one artificial radioactive decay chain of unstable heavy atomic nuclei that decay through a sequence of alpha and beta decays until a stable nucleus is achieved. Most radioisotopes do not decay directly to a stable state and all isotopes within the series decay in the same way. In physics of nuclear decays, the disintegrating nucleus is usually referred to as the parent nucleus and the nucleus remaining after the event as the daughter nucleus. Since alpha decay represents the disintegration of a parent nucleus to a daughter through the emission of the nucleus of a helium atom (which contains four nucleons), there are only four decay series. Within each series, therefore, the mass number of the members may be expressed as four times an appropriate integer (n) plus the constant for that series. As a result, the thorium series is known as the 4n series, the neptunium series as the 4n + 1 series, the uranium series as the 4n + 2 series and the actinium series as the 4n + 3 series.

Three of the sets are called natural or classical series. The fourth set, the neptunium series, is headed by neptunium-237. Its members are produced artificially by nuclear reactions and do not occur naturally.

- The thorium series (4n series),

- The uranium series (4n+2 series),
- The actinium series (4n+3 series),
- The neptunium series (4n+1 series).

The classical series are headed by primordial unstable nuclei. Primordial nuclides are nuclides found on the Earth that have existed in their current form since before Earth was formed. The previous four series consist of the radioisotopes that are the descendants of four heavy nuclei with long and very long half-lives:

- The thorium series with thorium-232 (with a half-life of 14.0 billion years),
- The uranium series with uranium-238 (which lives for 4.47 billion years),
- The actinium series with uranium-235 (with a half-life of 0.7 billion years),
- The neptunium series with neptunium-237 (with a half-life of 2 million years).

The half-lives of all the daughter nuclei are all extremely variable and it is difficult to represent a range of timescales going from individual seconds to billions of years. Since daughter radioisotopes have different half-lives then secular equilibrium is reached after some time. In the long decay chain for a naturally radioactive element, such as uranium-238, where all of the elements in the chain are in secular equilibrium, each of the descendants has built up to an equilibrium amount and all decay at the rate set by the original parent. If and when equilibrium is achieved, each successive daughter isotope is present in direct proportion to its half-life. Since its activity is inversely proportional to its half-life, each nuclide in the decay chain finally contributes as many individual transformations as the head of the chain.

As can be seen from figures, branching occurs in all four of the radioactive series. That means the decay of a given species may occur in more than one way. For example, in the thorium series, bismuth-212 decays partially by negative beta emission to polonium-212 and partially by alpha emission to thallium-206.

Radioactive cascade significantly influences radioactivity (disintegrations per second) of natural samples and natural materials. All the descendants are present, at least transiently, in any natural sample, whether metal, compound, or mineral. For example, pure uranium-238 is weakly radioactive (proportional to its long half-life), but a uranium ore is about 13 times more radioactive than the pure uranium-238 metal because of its daughter isotopes (e.g. radon, radium etc.) it contains. Not only are unstable radium isotopes significant radioactivity emitters, but as the next stage in the decay chain they also generate radon, a heavy, inert, naturally occurring radioactive gas. Radon itself is a radioactive noble gas, but the main issue is that it is a transport medium from bedrock to atmosphere (or inside buildings) for its short-lived decay products (Pb-210 and Po-210), that possess much more health risks.

Radiation from Uranium and its Decay Products

Uranium cascade significantly influences radioactivity (disintegrations per second) of natural samples and natural materials. All the descendants are present, at least transiently, in any natural uranium-containing sample, whether metal, compound, or mineral. For example, pure uranium-238 is weakly radioactive (proportional to its long half-life), but a uranium ore is about 13 times more radioactive than the pure uranium-238 metal because of its daughter isotopes (e.g. radon, radium etc.) it contains. Not only are unstable radium isotopes significant radioactivity emitters, but as the next stage in the decay chain they also generate radon, a heavy, inert, naturally occurring radioactive gas.

Radiation from Thorium and its Decay Products

Thorium cascade significantly influences radioactivity (disintegrations per second) of natural samples and natural materials. All the descendants are present, at least transiently, in any natural thorium-containing sample, whether metal, compound, or mineral. For example, pure thorium-232 is weakly radioactive (proportional to its long half-life), but a thorium ore is about 10 times more radioactive than the pure thorium-232 metal because of its daughter isotopes (e.g. radon, radium etc.) it contains. Not only are unstable radium isotopes significant radioactivity emitters, but as the next stage in the decay chain they also generate radon, a heavy, inert, naturally occurring radioactive gas.

parent nuclide	decay mode	half-life	daughter nuclide
232Th	α	1.405E10 a	228Ra
228Ra	β-	5.75 a	228Ac
228Ac	β-	6.25 h	228Th
228Th	α	1.91 a	224Ra
224Ra	α	3.63 d	220Rn
220Rn	α	55.6 s	216Po
216Po	α	0.145 s	212Pb
212Pb	β-	10.64 h	212Bi
212Bi	β- 64.06%	60.55 min	212Po
212Bi	α 35.94%		208Tl
212Po	α	299 ns	208Pb
208Tl	β-	3.053 min	208Pb
208Pb	stable		

Liquid Earth's Core

All three naturally-occurring isotopes of uranium (^{238}U, ^{235}U and ^{234}U) and naturally-occurring isotope of thorium have very long half-life (e.g. 4.47×10^9 years for ^{238}U). Because of this very long half-life uranium and thorium are weakly radioactive and contributes to low levels of natural background radiation in the environment. These isotopes are alpha radioactive (emitting alpha particle), but they can also rarely undergo a spontaneous fission.

All naturally-occurring isotopes belong to primordial nuclides, because their half-life is comparable to the age of the Earth (~4.54×10^9 years). Uranium has the second highest atomic mass of these primordial nuclides, lighter only than plutonium. Moreover the decay heat of uranium and thorium and their decay products (e.g. radon, radium etc.) contributes to heating of Earth's core. Together with potassium-40 in the Earth's mantle is thought that these elements are the main source of heat that keeps the Earth's core liquid.

Terrestrial Radiation

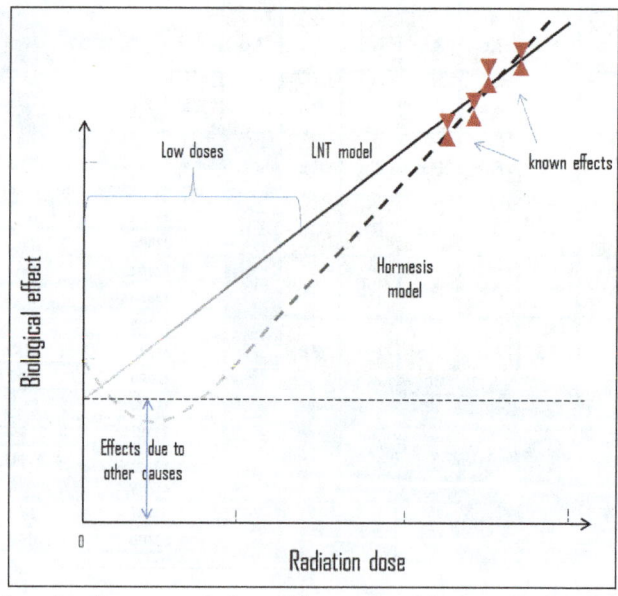

Alternative assumptions for the extrapolation of the cancer risk vs. radiation dose to low-dose levels, given a known risk at a high dose: LNT model, and hormesis model.

We must emphasize, eating bananas, working as airline flight crew or living in locations with, increases your annual dose rate. But it does not mean that it must be dangerous. In each case, intensity of radiation also matters. It is very similar as for heat from a fire (less energetic radiation). If you are too close, the intensity of heat radiation is high and you can get burned. If you are at the right distance, you can withstand there without any problems and moreover it is comfortable.

If you are too far from heat source, the insufficiency of heat can also hurt you. This analogy, in a certain sense, can be applied to radiation also from radiation sources.

In case of terrestrial radiation, we are talking usually about so called "low doses". Low dose here means additional small doses comparable to the normal background radiation (10 μSv = average daily dose received from natural background). The doses are very very low and therefore the probability of cancer induction could be almost negligible. Secondly, and this is crucial, the truth about low-dose radiation health effects still needs to be found. It is not exactly known, whether these low doses of radiation are detrimental or beneficial (and where the threshold is). Government and regulatory bodies assume a LNT model instead of a threshold or hormesis not because it is the more scientifically convincing, but because it is the more conservative estimate. Problem of this model is that it neglects a number of defence biological processes that may be crucial at low doses. The study during the last two decades is very interesting and shows that small doses of radiation given at a low dose rate stimulate the defense mechanisms. Therefore the LNT model is not universally accepted with some proposing an adaptive dose–response relationship where low doses are protective and high doses are detrimental. Many studies have contradicted the LNT model and many of these have shown adaptive response to low dose radiation resulting in reduced mutations and cancers. This phenomenon is known as radiation hormesis.

Types of Solar Radiation

Insolation is the downward solar energy flux at the ground surface in the shortwave region of electromagnetic spectrum. The solar radiation reaches earth's surface in two ways from extraterrestrial region:

- Beam radiation: A part of sun's radiation travels through earth's atmosphere and its reaches directly, which is called direct or beam radiation. The solar radiation along with the line joining the receiving point and the sun is called beam radiation. This radiation has any unique direction.

- Diffuse radiation: The remaining major part of the solar radiation is scattered, reflected back into the space or absorbed by earth's atmosphere. A part of this radiation may reach earth's surface. This radiation reaching earth's surface by the mechanism of scattering and reflecting, that is, radiation, is called diffuse or sky radiation. The solar radiation which is scattered by the particles in earth's atmosphere and this radiation dose not have any unique direction. The diffuse radiation takes pace uniformly in the all direction and its intensity does not change with the orientation of the surface.

Total or global radiation at any location on earth's surface = beam radiation + diffuse radiation.

However, direct or beam radiation depends on the orientation of the surface. The beam radiation depends on the angle of incident on the surface and its intensity is maximum when the solar radiation is falling normal to the surface. The solar radiation propagating normal to its direction is specified by I_n.

Air Mass (m)

The radiation reaching earth surfaces depend on (i) atmospheric condition and depletion and (ii) solar altitude. Air mass is the ratio of the path length through the atmosphere which the solar beam actually traverses up to earth's surface to the vertical path length through the atmosphere (minimum height of terrestrial atmosphere). At sea level, the air mass is unity when the sun is vertically is in the sky (inclination angle 90°).

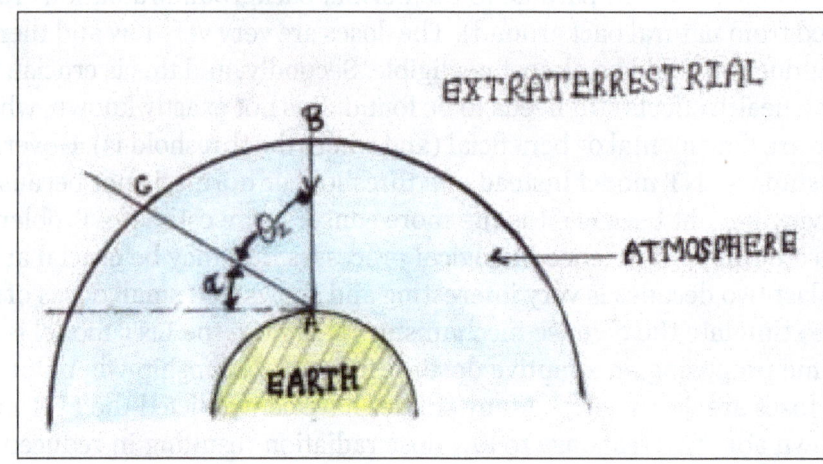

$$\text{Air mass} = \frac{\text{Path length travelled by beam radiation}}{\text{Vertical path length of the atmosphere}} = \frac{AC}{AB} = \text{cosec}\, \alpha = \sec \phi$$

Basic Sun-Earth Angles

Latitude or Angle of Latitude (λ)

The latitude location of earth's surface is the angle made by the radial line joining the specified location to the centre of earth with the projection of this line on the equatorial plane. The latitude equator is zero and it is 90° at poles. On a globe of the earth, lines of latitude are circles of different size. The largest one is the circle at equator (circle at equator with centre at earth' centre) whose latitude is taken as zero. The circles at the poles have latitude of 90° north and 90° south (or 90°) where these circles shrink to a point.

Longitude

On the globe, vertical lines of constant longitude (meridians) extend from pole to pole similar to the segment boundaries on peeled orange. Every meridian has to cross the equator and equator is circle. Like any circle, it has 360 degrees or division. Hence, longitude of a point is the marked value of that division where its meridian meets the equator circle. The meridian passing through the royal astronomical observatory at Greenwich, UK had been chosen as zero longitude. The meridian passing through this location is called prime meridian. The prime meridian or longitude is considered zero longitude and there are 180 longitude lines or degrees at cast (+180°) of Greenwich. The longitude lines meet at poles and these have wide separation at the equator (about 111 km). Solar noon is the time when the sun is at the longitude of the place.

Declination Angle (δ)

It is the angle made by the line joining the centres of sun and earth with equatorial plane.

Sun and Solar Energy

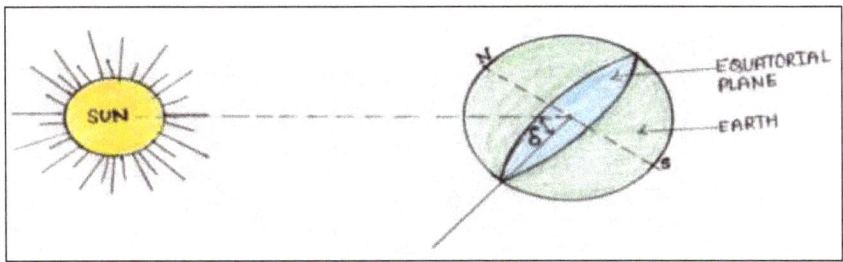

Declination angle.

The angle of declination varies when earth revolves around the sun. it has maximum value of 23.45° when earth achieves a position in its orbit corresponding to 21 June and it has minimum value of - 23.45° when earth is in orbital position corresponding to 22 December. The northern hemisphere. The angle of declination can be given by,

$$\delta = 23.45 \times \sin\left[\frac{360}{365}(284+n)\right]$$

Hour Angle (ω)

The hour angle at any instant is the angle through which the earth has to turn to bring the meridian of the observer directly in line with sun's rays. It is an angular measure of time. It is the angle in degree traced by the sun 1 h with reference to 12 noon of the (LAT) is positive in afternoon and negative in forenoon as shown in figure.

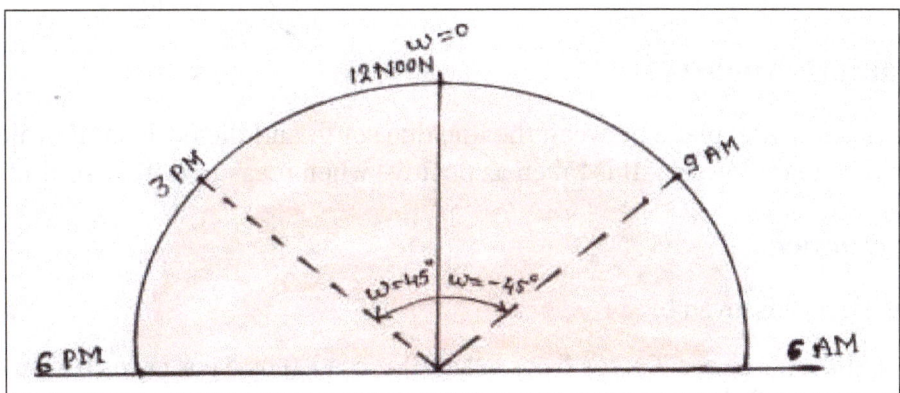

Hour angle.

The earth completes one rotation (360°) in 24 h. Hence, 1 h corresponds to 15° of earth rotation. As at solar noon the sun rays are in the line with local meridian or longitude, the hour angle at that instant is zero.

The hour angle can give as follows:

ω = [Solar time -12] x 15°

Zenith Angle (θ$_z$)

It is the angle between sun's ray and normal to horizontal plane as shown in figure.

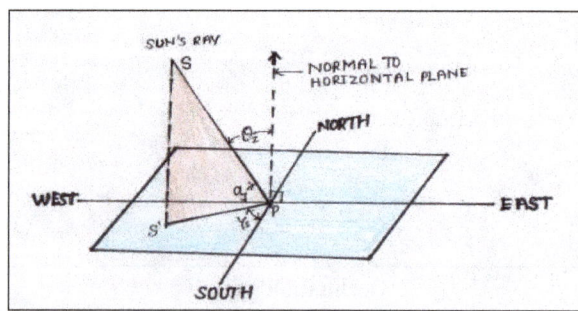

Zenith angle, solar azimuth angle and angle of incidence.

Solar Azimuth Angle (γ_s)

It is the angle between the projection of sun's ray to the point on the horizontal plane and line due to south passing through that point. The value of the azimuth angle is taken positive when it is measure from towards west.

Angle of Incidence (θ)

Angle of incidence for any surface is defined as the angle formed between the direction of the sun ray and the line normal to the surface as shown in the figure.

Tilt or Slope Angle (β)

The tilt angle is the angle between the inclined slope and the horizontal plane.

Surface Azimuth Angle (γ)

It is the angle is horizontal plane between the line due south and the horizontal projection of normal to the inclined plane surface. It is taken as positive when measured from south towards west.

Angle of Incidence

The angle of incidence is given by:

$$\cos \theta = \sin \lambda (\sin \delta \cos \beta + \cos \delta \cos \omega \sin \beta) + \cos \lambda (\cos \delta \cos \omega \cos \beta - \sin \delta \cos \gamma \sin \beta) + \cos \delta \sin \gamma \sin \beta \sin \omega$$

Local Apparent Time

LAT is the time used for determining hour angle.

LAT = standard time + 4 (standard time longitude − location longitude) + (equation of time correction)

Calculation of LAT

Example: Calculate hour angle for local apparent time given by:

- 6 AM

Sun and Solar Energy

- 9 AM
- 12 noon
- 3 PM
- 6 PM vi) 9 PM

Solution: The hour angle is given by ω = [Solar time -12] x 15°

- Solar time 6 AM ω = [6 -12] x 15° = -90°
- Solar time 9 AM ω = [9 -12] x 15° = -45°
- Solar time 12 noon ω = [12 -12] x 15° = 0°
- Solar time 3 PM= 15 ω = [15 -12] x 15° = 45°
- Solar time 6 PM =18 ω = [18 -12] x 15° = 90°
- Solar time 9 PM=21 ω = [21 -12] x 15° = 135°

Solar Day Length

The angle between sunrise and sunset is the solar day length. It is given by:

Solar day length = -2 ω = 2 cos-1 (-tan λ tan δ)

The duration of sunshine hour (t_d) or daylight hours is given by:

$$t_d = \frac{\text{Angle between sunrise and sunset}}{15°} \times 1h$$

$$= \frac{2\omega}{15} \times 1 = \frac{2\cos^{-1}(-\tan\lambda\tan\delta)}{15}$$

Examples

1. Calculate the number of daylight hours in Srinagar (latitude of 34°05') for 1 January and 1 July?

Solution:

- Declination angle,

$$\delta = 23.45 \times \sin\left[\frac{360}{365}(284+n)\right]$$

Where, n is number of days of the year counted from 1 January.

- Case1: 1 January

  n = 1

$$\delta = 23.45 \times \sin\left[\frac{360}{365}(284+1)\right]$$

$$= -23.01°$$

$$t_d = \frac{2\cos^{-1}(-\tan\lambda\tan\delta)}{15} = \frac{2\cos^{-1}(-\tan 34.08\tan 23.01°)}{15} = 9.77\,h$$

- Case 2: 1 July

    n=182

$$\delta = 23.45 \times \sin\left[\frac{360}{365}(284+182)\right]$$

$$= -23.12°$$

$$t_d = \frac{2\cos^{-1}(-\tan\lambda\tan\delta)}{15} = \frac{2\cos^{-1}(-\tan 34.08\tan 23.12°)}{15} = 14.24\,h$$

2. Calculate the day length in hours in Mumbai on 1 December if latitude is 19.116 and declination angle is -22.108°.

$$t_d = \frac{2}{15}\cos^{-1}(-\tan\lambda\tan\delta)$$

$$= \frac{2}{15}\cos^{-1}(-\tan 19.116 \times \tan-22.108°)$$

$$= \frac{2}{15} \times 81.92 = 10.9\,h$$

3. Find the solar altitude at 2 h after local solar angle on 1 June for place located at 26.750 N latitude. Determine sunrise and sunset hours as well as day length?

Solution:

- n = 31 + 28 + 31 + 30 + 31 + 1 = 152

$$\delta = 23.45 \times \sin\left[\frac{360}{365}(284+n)\right] = 23.45 \times \sin\left[\frac{360}{365}(284+152)\right] = 22°$$

- Hour angle, ω = 2h x 15° = 30°

- $\sin\alpha = \sin\lambda \cdot \sin\delta + \cos\lambda\cos\delta\cos\omega$

    = sin 26.75 x sin 22 + cos 26.75 x cos 22 x cos 30 = 0.553

    α = 72.4°

- $t_d = \frac{2}{15}\cos^{-1}(-\tan 26.75 \times \tan 22) = 10.40\,h$

- The sunrise is at (12 - 10.4/2)h = (12 - 5.2) = 6.8 h = 6:48 AM
- The sunset is at (12 + 5.2)h = 17.2 h = 5:12 PM

Intensity of Terrestrial Radiation

The normal intensity (I_N) is the intensity normal to horizontal surface, beam radiation (I_b) and diffuse radiation (I_d) on horizontal surface is given by:

$$I_b = I_N \cos \theta z$$

$$I_d = 1/3 (I_{ext} - I_N) \cos \theta z$$

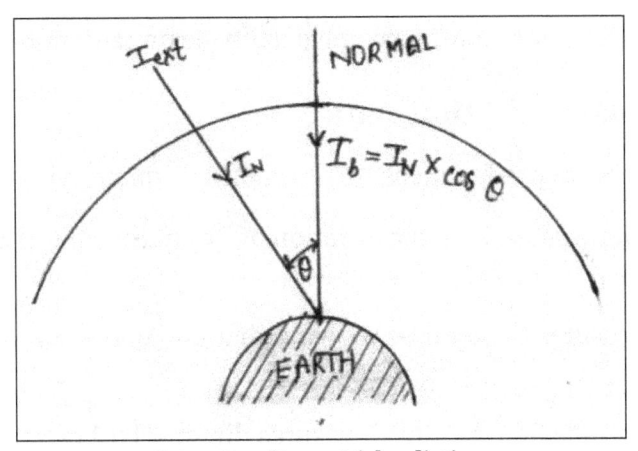

Intensity of terrestrial radiation.

Total or global radiation:

$$I_G = I_b \times R_b + I_d \times R_d + R_r (I_b \times I_d)$$

Where, R_b, R_d, R_r are conversion factors for beam, diffuse and reflected components.

Calculating Solar Radiation Data

The solar radiation is estimated as follows:

- Monthly average daily global radiation, H_g

$$\frac{\overline{H_g}}{\overline{H_o}} = a + b(\overline{n}/\overline{N})$$

Where,

- H_o : Monthly average daily extraterrestrial radiation.
- $\overline{n}$: Monthly average daily hours of sunshine.
- $\overline{N}$: Monthly average maximum possible hours of sunshine.
- a and b are regression parameters (constant) for a location.

- Monthly average daily diffuse radiation H_o on horizontal surface

$$\frac{\overline{H_d}}{\overline{H_g}} = 1.194 - 0.838 \overline{k_T} - 0.0446 (\overline{n}/\overline{N})$$

Where, KT: Monthly average clarity index.

- Monthly average daily global radiation on tilted surface:

$$\overline{H_T} = (\overline{H_g} - \overline{H_d})\overline{R_b} + \overline{H_d}\frac{(1+\cos\beta)}{2} + \overline{H_g}\rho \times \frac{(1+\cos\beta)}{2}$$

Where, Rb: Conversion factor for monthly average dailty beam radiation.

Measurement of Solar Radiation Data

The following instruments are used to measure duration and quantifying solar radiation:

- Sunshine Recorder: It measures the duration of Bright Sunshine (BSS) Hours in any given day.

- Pyranometer: Measures the total or Global radiation i.e. the summation of direct and diffuse radiation.

- Pyrheliometer or Tube Solarimeter: Measures direct solar beam i.e. radiation falling on a plane surface at normal (perpendicular) incidence angle.

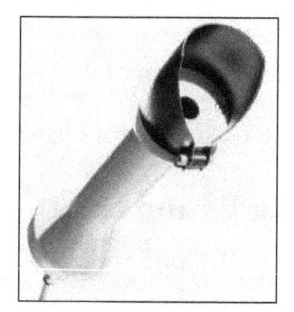

a) Pyranometer.   b) Pyrheliometer.

Pyranometer

Pyranometer is a device that measures solar irradiance from a hemispherical field of view incident on a flat surface. The SI units of irradiance are watts per square metre (W/m²). Traditionally pyranometers were mainly used for climatological research and weather monitoring purposes, however recent worldwide interest in solar energy has also lead to an increased interest in pyranometers.

Pyranometers measure global irradiance: the amount of solar energy per unit area per unit time incident on a surface of specific orientation emanating from a hemispherical field of view (2π sr), denoted Eg↓. The global irradiance includes direct sunlight and diffuse sunlight, as illustrated in Figure. The contribution from direct sunlight is given by E·cos (θ) where θ is the angle between the surface normal and the position of the sun in the sky and E is the maximum amount of direct sunlight. The global irradiance is then:

$$Eg{\downarrow} = E \cdot \cos(\theta) + Ed$$

Where, Ed accounts for the diffuse sunlight.

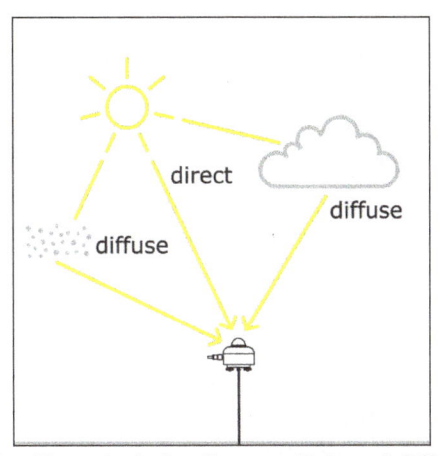

The global irradiance includes direct sunlight and diffuse sunlight.

In many cases the surface of interest is horizontal such that the hemispherical field of view corresponds to the sky dome. In that case the measured quantity is the so called global horizontal irradiance (GHI) denoted $Eg{\downarrow}h$. In some cases the surface is tilted, for example in photovoltaic applications where the surface often corresponds to the plane of array (POA) of solar panels. In this case the measured quantity is the global tilted irradiance (GTI) denoted $Eg{\downarrow}t$.

A special case is the case were the surface is horizontal, but with the pyranometer facing downwards instead of towards the sky. In this case the measured quantity is the diffuse reflection from the surface of the earth, denoted $Er{\uparrow}$.

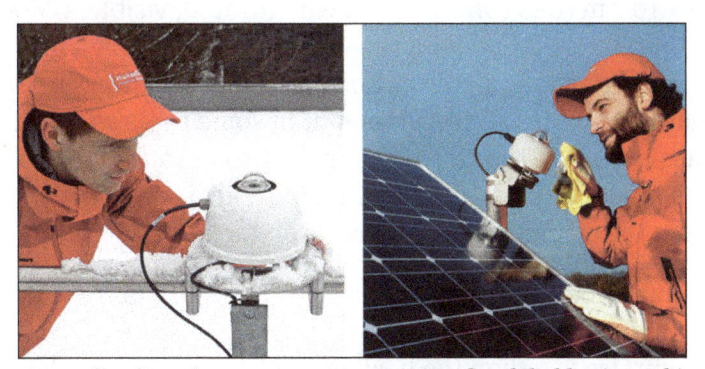

Left: a horizontally aligned pyranometer measuring the global horizontal irradiance (GHI) and right: a tilted pyranometer measuring a global tilted irradiance (GTI).

The global irradiance may vary greatly depending on the height of the sun in the sky (and thus location on the earth, time of day and time of year) and on meteorological and environmental factors such as clouds, aerosols, smog, fog, precipitation and others. Typical values for the global horizontal irradiance are in the range from 0 to 1400 W/m². In some cases it can be larger for example due to reflections from buildings or snow or in a more exotic example at the centre of a solar concentrator.

Use of Pyranometer

The sun is earth's main source of extraterrestrial energy. This has important implications in two areas: weather and climate on the one hand and energy production by harvesting solar energy on the other hand.

Solar radiation is one of the driving forces behind the earth's weather patterns and thus an important factor in weather and climate studies. In such studies pyranometers are mostly used to measure the GHI to determine the irradiance incident on the surface of the earth. The GHI that one would measure just outside earth's atmosphere is fairly predictable, but at the surface of the earth the irradiance depends strongly on factors such as cloud coverage, aerosol concentration, fog and smog. Another interesting measurement is that of the net irradiance $E^* = Eg\downarrow - Er\uparrow$ or the albedo $A = Er\uparrow/Eg\downarrow$. In this case two horizontally aligned pyranometers are used: one facing towards the ground and one facing towards the sky.

In the solar energy industry pyranometers are used to monitor the performance of photovoltaic (PV) power plants. By comparing the actual power output from the PV power plant to the expected output based on a pyranometer reading the efficiency of the PV power plant can be determined. Drops in efficiency may indicate that maintenance of the PV plant is required. Pyranometers can also be used to determine the suitability of potential sites for PV power plants. In this case pyranometers are used to determine the expected output of a PV installation.

Working Principles of Pyranometer

Pyranometers are irradiance sensors that are based on the Seebeck- or thermoelectric effect. The main components of a pyranometer are one or two domes, a black absorber, a thermopile, the pyranometer body and in some cases additional electronics.

The dome on a pyranometer acts as a filter that transmits solar radiation with wavelengths from roughly 0.3 to about 3×10^{-6} m (this contains the near-infrared, visible, UV-A and part of the UV-B radiation), but blocks thermal radiation with wavelengths longer than 3 μm. Occasionally a second dome is used to improve the pyranometer performance. Pyranometer domes are typically made from Schott N-BK7 glass or Schott WG295 glass, but in some cases sapphire or fused silica (Spectrosil or Infrasil) domes are used. The transmission τ of solar radiation through a dome is ideally close to 100 %, but is in practice closer to 92 %. The dome also serves to protect the black absorber and the thermopile from the elements (rain, snow, etc.).

The filtered radiation is absorbed by the black surface on the pyranometer and converted into heat. If the transmission through the dome(s) is τ, the area of the black surface is A and the absorption coefficient of the black surface is α then the heat absorption can be calculated as follows:

$$P_{absorption} = \alpha \cdot \tau \cdot A \cdot Eg\downarrow$$

This creates a temperature gradient from the black surface through the thermopile to the pyranometer body which acts as a heatsink. The temperature difference is given by:

$$\Delta T = R_{thermal} \cdot P_{absorption}$$

Where, Rthermal is the thermal resistance of the thermopile sensor. This thermal resistance depends on the specific composition and geometry of the thermopile sensor. A thermopile consists of a number of thermocouples connected in series. Each thermocouple will generate a voltage proportional to the temperature difference between the black surface and the body:

$$u = \varsigma \cdot \Delta T$$

Where, ς is the Seebeck coefficient. For example, the Seebeck coefficient of a copper-constantan thermocouple is 41×10^{-6} V/K.

Types of Pyranometer

Pyrometers are classified into two types like thermopile pyranometer, photodiode-based pyranometer.

Thermopile Pyranometer

A thermopile pyranometer (also called thermo-electric pyranometer) is a sensor based on thermopiles designed to measure the broad band of the solar radiation flux density from a 180° field of view angle. A thermopile pyranometer thus usually measures 300 to 2800 nm with a largely flat spectral sensitivity. The first generation of thermopile pyranometers had the active part of the sensor equally divided in black and white sectors. Irradiation was calculated from the differential measure between the temperature of the black sectors, exposed to the sun, and the temperature of the white sectors, sectors not exposed to the sun or better said in the shades.

In all thermopile technology, irradiation is proportional to the difference between the temperature of the sun exposed area and the temperature of the shadow area.

Design

In order to attain the proper directional and spectral characteristics, a thermopile pyranometer is constructed with the following main components:

- A thermopile sensor with a black coating: It absorbs all solar radiation, has a flat spectrum covering the 300 to 50,000 nanometer range, and has a near-perfect cosine response.
- A glass dome: It limits the spectral response from 300 to 2,800 nanometers (cutting off the part above 2,800 nm), while preserving the 180° field of view. It also shields the thermopile sensor from convection. Many, but not all, first-class and secondary standard pyranometers include a second glass dome as an additional "radiation shield", resulting in a better thermal equilibrium between the sensor and inner dome, compared to some single dome models by the same manufacturer. The effect of having a second dome, in these cases, is a strong reduction of instrument offsets. Class A, single dome models, with low zero-offset (+/- 1 W/m²) is available.

In the modern thermopile pyranometers the active (hot) junctions of the thermopile are located beneath the black coating surface and are heated by the radiation absorbed from the black coating. The passive (cold) junctions of the thermopile are fully protected from solar radiation and in thermal contact with the pyranometer housing, which serves as a heat-sink. This prevents any alteration from yellowing or decay when measuring the temperature in the shade, thus impairing the measure of the solar irradiance.

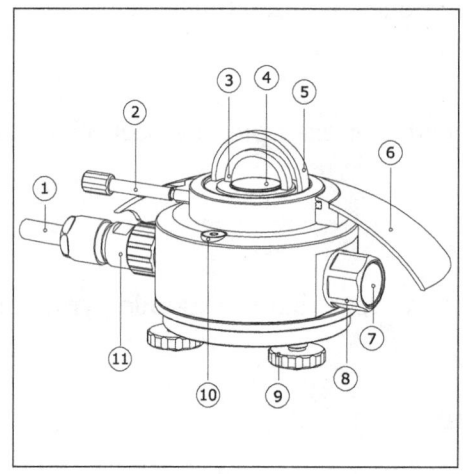

Linedrawing of a pyranometer, showing essential parts: (1) cable, (3)pyranameter and (5) glass domes, (4) black detector surface, (6) sun screen, (7) desiccant indicator, (9) levelling feet, (10) bubble level, (11) connector.

The thermopile generates a small voltage in proportion to the temperature difference between the black coating surface and the instrument housing. This is of the order of 10 μV (microvolts) per W/m^2, so on a sunny day the output will be around 10 mV (millivolts). Each pyranometer has a unique sensitivity, unless otherwise equipped with electronics for signal calibration.

Working Principle of Thermopile Pyranometer

The thermoelectric detection principle is used, whereby incoming radiation is almost completely absorbed by a horizontal blackened surface, over a very wide wavelength range. The resulting increase of temperature is measured via thermocouples connected in series or series-parallel to make a thermopile.

The active (hot) junctions are located beneath the blackened receiver surface and are heated by the radiation absorbed in the black coating. The passive (cold) junctions of the thermopile are in thermal contact with the pyranometer housing, which serves as a heat-sink. More recent, higher performance, pyranometers use a Peltier element. This is also thermoelectric, but the dissimilar metals of a thermocouple / thermopile are replaced by dissimilar semiconductors.

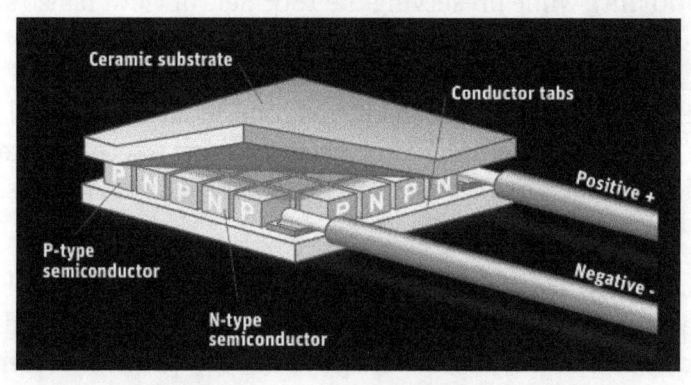

It is necessary to protect the black detector coating against external influences which may affect the measurement; such as precipitation, dirt and wind. Nearly all pyranometers use an optical quality glass for their hemispherical single or double domes.

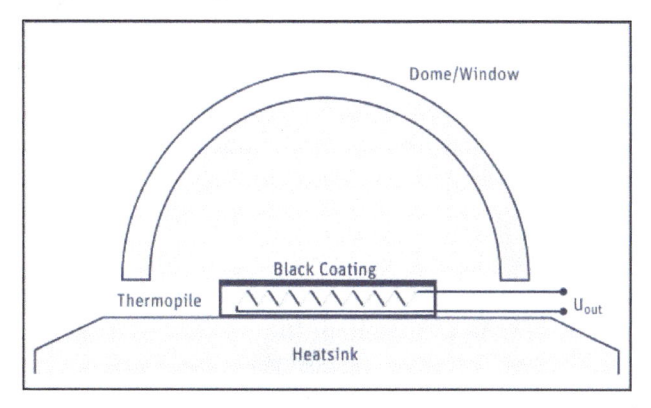

Depending upon the glass, the transmission is from 300 nm or less to about 3000 nm. Double domes give better stability under dynamically changing conditions by further 'insulating' the sensor surface from environmental effects such as wind and rapid temperature fluctuations.

The shape of the dome, and the refractive index of the material, improves the response of the sensor when the sun is close to the horizon, 'bending' the incoming radiation beam. The highest specification pyranometer available, our modelCMP 22 uses Quartz domes for a wider spectral response. The higher refractive index further improves the directional response and better thermal conductivity than glass provides other performance benefits.

Passive thermopile / Peltier pyranometers such as our CMP series do not require a power supply. The detector generates a small voltage in proportion to the temperature difference between the black absorbing surface and the instrument housing. This is of the order of 10 μV (microvolts) per W/m^2, so on a sunny day the output will be around 10 mV (millivolts). Each pyranometer has a unique sensitivity, defined during the calibration process, which is used to convert the output signal in microvolts into global irradiance in W/m^2.

Our SMP ranges of Smart pyranometers have the same detectors as the equivalent CMP models but with built-in digital signal processing and performance enhancement, and therefore they require external power to operate.

To maintain performance, recalibration is usually recommended every two years, and a high quality water-proof connector for the signal cable greatly simplifies the process.

Photodiode-based Pyranometer

Photodiode-based pyranometer can be used in any installation where reliable measurement of solar irradiance is necessary, especially in those where cost may be a deciding factor.

Generically, a pyranometer is a device for measuring solar radiation on a normally flat surface, in a field of 180 degrees. Measurement of solar radiation per unit of surface (W = m²) is termed irradiance. Irradiance measurement requires, by definition, that the pyranometers sensors response to radiation varies with the cosine of the angle of incidence from a line vertical to the surface of the sensor. The difference between the pyranometers real response and the ideal cosine response is termed cosine error. Pyranometers are widely used in passive solar systems analysis, meteorology studies, climatology, agriculture, irrigation scheduling, solar energy studies and building physics. In spite of the interest in measuring solar radiation, the use of pyranometers is still not very widespread outside the field of research, probably due to their high cost.

Comparison between References Cell and Pyranometer

Reference cells show similar properties to PV panels, but even after the process of calibration, they have similar shortcomings in temperature and spectrum range. Therefore, they will not be able to give an accurate measurement of the available solar radiation under all conditions. A pyranometer has the following advantages over reference cell:

- The pyranometer gives an independent, ac-curate reading of the total available solar radiation.
- The pyranometer gives an independent, ac-curate reading of the total available solar radiation.
- The pyranometer are classified and calibrated to ISO standards.
- The response time of the pyranometer is longer than a PV cell.
- The pyranometer is PV cell type independent.
- A pyranometer can have a very small temperature coefficient.
- PV cells are specified at STC (Standard Test Conditions).
- Reference cells (and PV panels) suffer more from pollution than pyranometers.
- Performance Ratio or Performance Index calculations are more accurate using a pyranometer.

Characteristics of Pyranometer

The element that characterizes a pyranometer is the sensor it uses, which may be thermal (thermopile) or photovoltaic. Photovoltaic sensors are a cheap alternative, whose only advantage in principle over thermopiles in measuring radiation, aside from their price, is their response speed. Thus, while photodiode-based pyranometers have a response time of around 10μs, in those based on thermopiles, response time ranges between 1 and 10s making them less suitable for measuring very rapid changes in radiation. The influence of temperature on pyranometer 0s measurement is also well known. Although this influence exists, it is lower in thermopile pyranometers than in photodiode devices. With regard to integrating a pyranometer into an instrumentation system (generally into any measuring device), there is a series of very important factors to take into consideration, namely:

- Ease of connection.

- Signal degradation due to the transmission process.

Features of the Proposed Photodiode based Pyranometer

In order to achieve the objective proposed in this work, designing and building a photodiode-based pyranometer with similar characteristics to those of a thermopile-based device, also incorporating significant connection, measuring and programming utilities, the authors have analyzed and corrected both the defects mentioned in literature and those observed during the testing of various commercial units. That is, the pyranometer developed has the following original features:

- Excellent cosine response guaranteed by both the level gauge (to guarantee horizontality), which is incorporated, and by the specifically designed solar radiation diffuser. Insensitivity in measuring variations in ambient temperature. A control circuit keeps temperature constant in the interior of the device.

- Its interior incorporates all necessary electronics for both conditioning and controlling which minimizes noise and the need for auxiliary electronics.

- Connection features in the proposed pyranometer are significant, both in terms of their quality (ease, robustness, immunity to noise, etc.) and the cost-saving involved in not having to transmit and condition analogue signals outside the device.

- In order to avoid internal condensation due to the temperature and air-tightness of the device which may degrade its electronic circuitry and steam up the lens of the photodiode sensor the proposed pyranometer must be equipped with a hygroscopic-salts container.

- The cost will be tens of times cheaper than that of a thermopile-based pyranometer of similar quality (including all signal conditioning).

Solar Angle of Incidence

The irradiance sensors response to the direct (beam) irradiance component is influenced by the cosine of the solar angle-of-incidence (AOI), and by the optical characteristics of its front surface. The response of the sensor to diffuse irradiance can be assumed to have no dependence on angle-of-incidence. The optical influence of the front surface could be a flator domed-glass cover or a translucent diffuser. To make a measurement of irradiance, it is required by definition that the response to beam radiation varies with the cosine of the angle of incidence, so that there will be a full response when the solar radiation hits the sensor perpendicularly (normal to the surface, sun at zenith, 0 degrees angle of incidence), zero response when the sun is at the horizon (90 degrees angle of incidence, 90 degrees zenith angle), and 0.5 at 60 degrees angle of incidence. Therefore, it can be deduced from the definition that a pyranometer must have a directional response or, as it is usually termed, a co-sine response to emphasize the fact that its response must ideally be analogous to the cosine function. Fig. illustrates the relative response of the irradiance sensors versus the solar angle-of-incidence. The sensors with a planar glass front surface have a stronger sensitivity to AOI, for angles greater than 60 degrees. To some degree, the stronger sensitivity is offset by the observation that the planar devices have more repeatable behavior, device to device, than many commercial pyranometers. Users should recognize that all pyranometers are subject to significant measurement errors at high AOI due to mechanical misalignment. For instance at AOI = 70 degrees, mounting a pyranometer only 1 degree different from the plane of a photovoltaic array will result in a 5% error in measured irradiance.

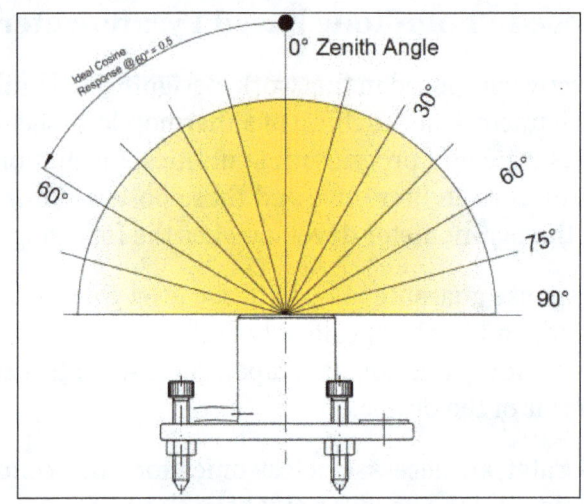

Cosine Response of Pyranometer.

Block Diagram

Teflon Diffuser

The Teflon Diffuser eliminates the cosine error to a large extent. Teflon has been used because it is a good diffuser and is also resistant to the elements and ultra-violet (UV) radiation, given its capability to diffuse transmitting lights nearly perfectly. Moreover, the optical properties of PTFE (TeflonTM) remain constant over a wide range of wavelengths, from UV up to near infrared.

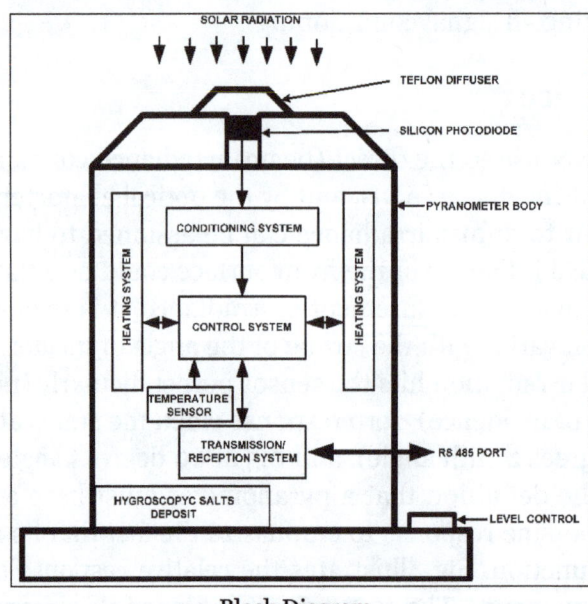

Block Diagram.

Most commercial pyranometers use a glass dome which, apart from being more expensive than the TeflonTM diffuser used in this pyranometer, becomes affected by continuous solar radiation and traps higher amounts of dirt.

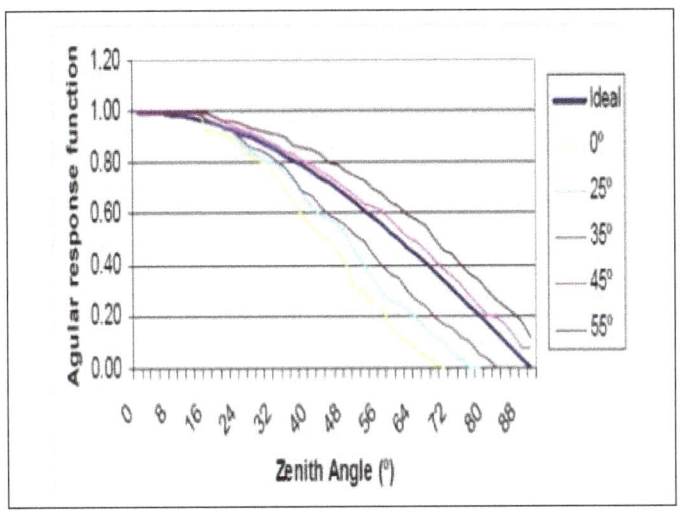

Deviation from ideal cosine response for different angles in the machined in the TeflonTM diffuser.

Pyranometer Housing

The pyranometer housing contains the photodiode and all the signal conditioning and distribution electronics. It is manufactured from a single piece of 10 mm thick black polyethylene, since polyethylene is a material which resists the elements very well and also shows excellent characteristics as a thermal insulator.

Hygroscopic Salts Deposit

To avoid condensation inside the pyranometer hygroscopic salts container is used. Use of these hygroscopic salts will ensure moisture levels to a minimum. Because of their affinity for atmospheric moisture, hygroscopic materials necessitate their being stored in sealed containers.

Sensor

The choice of pyranometer sensor element (photodiode) has required an exhaustive study of the commercial devices available, since it constitutes one of the key elements to being able to obtain better performance from the developed pyranometer. A photodiode was required with a response within the visible spectrum, a high value and as linear as possible. Using the characteristics in the datasheets supplied by the manufacturers, the great varieties of photodiodes analyzed were classified into two types, namely, those which incorporate the conditioning circuit and those which do not. The former were rejected immediately, as they exhibited problems of saturation at high luminosity. As for the latter, the following were analyzed: BPW21, OSD5-5T, OSD15-5T and S9219-01. In the datasheets for each photodiode the following characteristics were studied:

- Radiant sensitive area (mm^2) and spectral sensitivity (A/W): For a given irradiance (W/m^2), these two characteristics allow the level of the signal provided by the photodiode to be known.

- Noise equivalent Power (W/Hz1=2): Based on spectral sensitivity, this characteristic allows the noise-signal to be calculated. The signal produced by the photodiode divided by the noise signal is its signal-to-noise ratio (SNR).

- Price: The prices of the aforementioned photodiodes range from 7 to 20 Euros, with the cheapest being the BPW21. After the previous analysis, a practical test on the four photodiodes mentioned was carried out in the laboratory. For this, the following experiment was prepared to measure the voltage of each photodiode in short-circuit at different levels of irradiance.

Proposed Circuit

Conditioning System

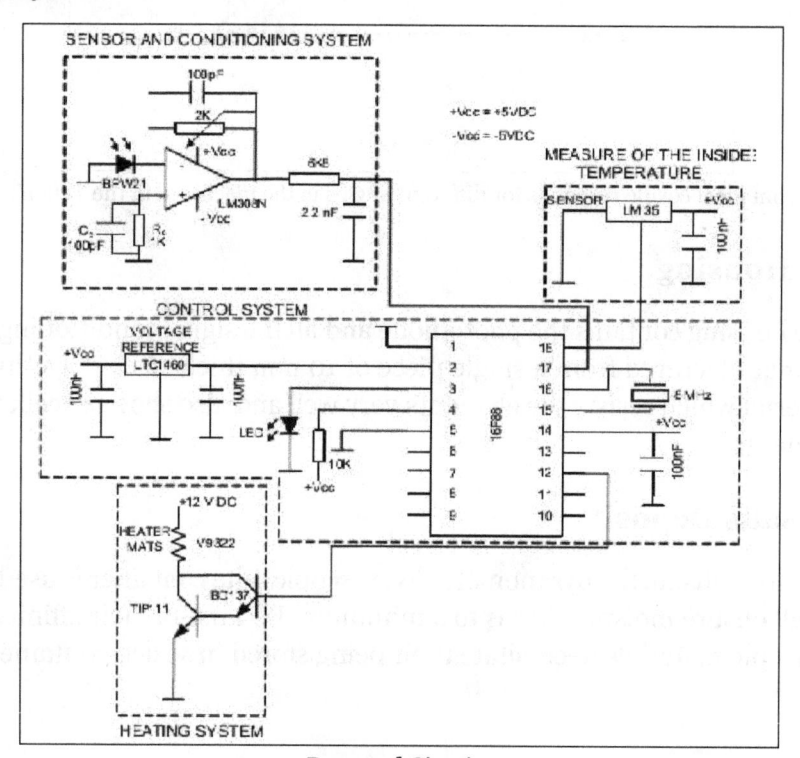

Proposed Circuit.

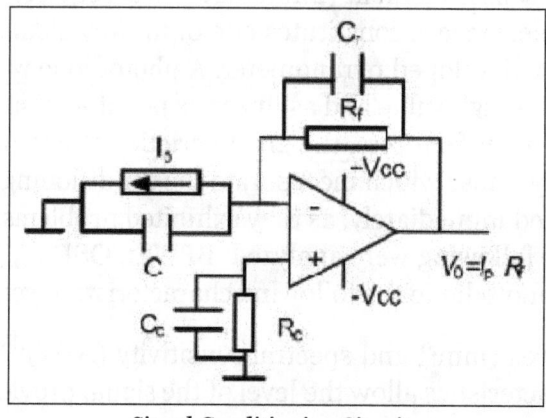

Signal Conditioning Circuit.

To calculate the value of Rf a nominal irradiance of 1,000 W/m² is used. For this, the BPW21 photodiode produces the photocurrent I_p = 2.49 x 10 3 A. Therefore, as the maximum analogue input value accepted by the Analogue-to-Digital Converter (ADC) is 2.5 V, the value of Rf is 1K, which is implemented using a 2 K multi-turn potentiometer to carry out precise adjustment. In order to correct the DC error due to polarization currents, a resistor (Rc) is connected to the non-inverting input of the OPAM. This resistor has a detrimental effect in terms of noise, which is amplified; this is why a 100 pF compensation capacitor Cc is connected in parallel with it. The parasitic capacitor on the photodiode BPW21, C, is 580 pF. This capacitor has to be taken into consideration, as it can influence the stability of the assembly (reducing its phase margin, and therefore, its relative stability). To improve the stability of the amplifier a capacitor Cr is connected in parallel with the feedback resistor Rf. Following the procedure laid down in the bibliography it is calculated that an appropriate value for the capacitor is 100 pF. Finally, a low-pass filter is connected to the amplifier output set at the frequency of 10 Hz (R = 6K8 and C = 2.2 F). In this way the possible interference that could affect the ADC input is minimized.

Control System

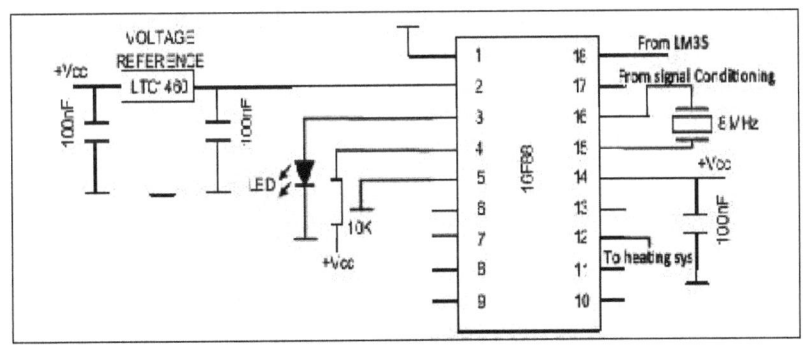

Controller Circuit.

A PIC-type microcontroller is used to control the entire pyranometer. The integrated circuit (IC) selected is 16F88, which incorporates an ADC. The ADC in the PIC acquires the conditioned analogue signal from the photodiode and converts it into digital format. The PIC also maintains the inside of the pyranometer at a constant temperature. For this reason, it receives the signal from an analogue temperature sensor: LM35 (chosen for its stability and precision), fitted in the interior of the pyranometer housing.

Heating System

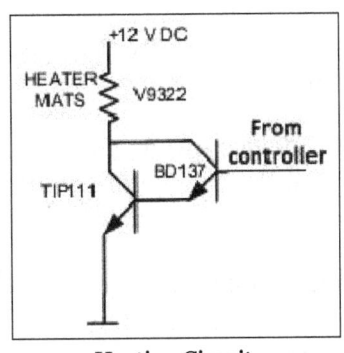

Heating Circuit.

Its job is to keep the temperature in the interior of the pyranometer constant at all times. Based on the operating temperature set by the user, the control system sends a signal to the thermostatisation system to activate the heaters until this temperature is reached.

The heaters are heating meshes (circular elements) which run on 12 V with an approximate current consumption of 400 mA. Logically, the control signal from the PIC is not applied directly to the heaters, but to an electronic power stage, made up of BD137 and TIP 111 transistors. The total power consumption of the device depends on the exterior temperature. However, the system is highly optimised, since the body of the pyranometer, made from 10 mm thick polyethylene, acts as an excellent thermal insulator. From the pyranometer control software, the user can select the minimum level of irradiance for the heating system to operate. This allows, for example, the pyranometer to stop working automatically at night and to start working, also automatically, by day. This utility allows optimisation of energy costs.

Pyrheliometer

The pyrheliometer is one type of instrument, used to measure the direct beam of solar radiation at the regular occurrence. This instrument is used with a tracking mechanism to follow the sun continuously. It is responsive to wavelengths bands that range from 280 nm to 3000 nm. The units of irradiance are W/m².

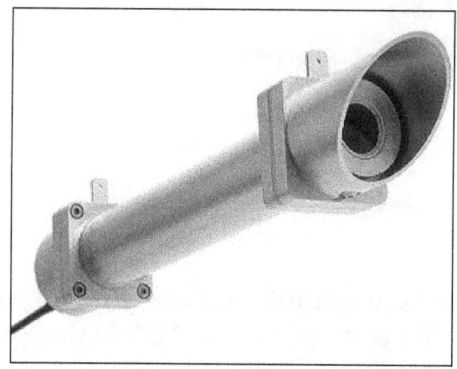

Pyrheliometer Instrument.

Pyrheliometer Construction and Working Principle

The external structure of the Pyrheliometer instrument looks like a telescope because it is a lengthy tube. By using this tube, we can spot the lens toward the sun to calculate the radiance. The Pyrheliometer basic structure is shown below. Here the lens can be pointed in the direction of the sun & the solar radiation will flow throughout the lens, after that tube & finally at the last part where the last apart includes a black object at the bottom.

The irradiance of solar enters into this device through a crystal quartz window and directly reaches onto a thermopile. So this energy can be changed from heat to an electrical signal that can be recorded.

A calibration factor can be applied once changing the mV signal to a corresponding radiant energy flux, and it is calculated in W/m² (watts per square meter). This kind of information can be used to increase Insolation maps. It a solar energy measurement, that is received on a specified surface region in a specified time to change around the Globe. The isolation factor for a specific area is very useful once setting up solar panels.

Pyrheliometer Circuit Diagram

The circuit diagram of the pyrheliometer is shown below. It includes two equal strips specified with two strips S1 & S2 with area 'A'. Here, a thermocouple is used where it's one junction can be connected to S1 whereas the other is connected to S2. A responsive galvanometer can be connected to the thermocouple. The S2 Strip is connected to an exterior electrical circuit.

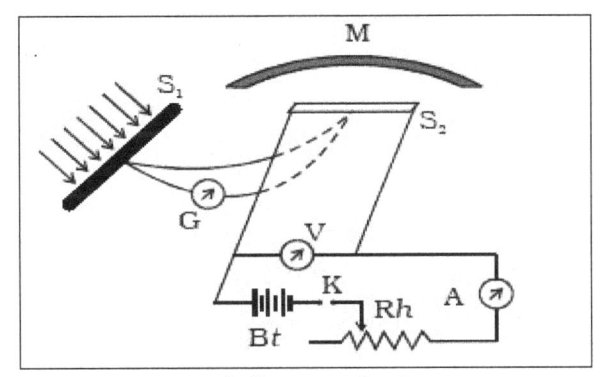

Pyrheliometer Circuit.

Once both the strips are protected from the radiation of solar, then the galvanometer illustrates there is no deflection because both the junctions are at equal temperature. Now 'S1' strip is exposed to the solar radiation & S2 is protected with a cover like M. When S1 strip gets heat radiations from the sun, then strip temperature will be increased, thus the galvanometer illustrates deflection.

When current is supplied throughout the S2 strip, then it is adjusted and the galvanometer illustrates there is no deflection. Now, again both the strips are at equal temperature.

If the heat radiation amount occurred over the unit area within the unit time on S1 strip is 'Q' & its absorption co-efficient, so the heat radiation amount which is absorbed through the S1 strip S1 within unit time is 'QAa'. In addition, the heat generated in unit time within the S2 strip can be given through VI. Here, 'V' is the potential difference & 'I' is the flow of current through it.

When heat absorbed is equivalent to the heat generated, so,

$$QAa = VI$$

$$Q = VI/Aa$$

By substituting the values of V, I, A and a, the value of 'Q' can be calculated.

Different Types

There are two types of Pyrheliometers like SHP1 and CHP1.

SHP1

The SHP1 type is a better version compare with CHP1 type, as it is designed with an interface including both improved analog o/p & digital RS-485 Modbus. The response time of this kind of meter has below 2 seconds & independently calculated temperature correction will range from -40 °C to + 70 °C.

CHP1

The CHP1 type is the most frequently used radiometer used to measure solar radiation directly. This meter includes one thermopile detector as well as two temperature sensors. It generates an utmost o/p like 25mV beneath usual atmospheric situations. This type of device totally obeys the most recent standards which are set by ISO and WMO about the criteria of the Pyrheliometer.

Advantages

The advantages of the Pyrheliometer include the following:

- Very low power consumption,
- Operates from a wide range of voltage supplies,
- Ruggedness,
- Stability.

Pyrheliometer Applications

The applications of this instrument include the following:

- Scientific meteorological,
- Observations of Climate,
- Testing research of Material,
- Estimation of the solar collector's efficiency,
- PV devices.

Difference between Pyrheliometer and Pyranometer

Both the instruments like Pyrheliometer & Pyranometer are used to calculate solar irradiance. These are related in their intention but there are some dissimilarities in their construction & working principle.

Pyranometer	Pyrheliometer
It is one kind of acidometer mainly used to measure the solar irradiance over a planar surface.	This instrument is used to measure direct ray solar irradiance.
It uses thermoelectric detection principle.	In this, the thermoelectric detection principle is used.
In this, the measurement of increasing temperature can be done through thermocouples which are linked in series otherwise series-parallel to build a thermopile.	In this, the increasing temperature can be calculated through thermocouples that are allied in series/series-parallel to create a thermopile.

This is frequently used in meteorological research stations.	This is also used in meteorological research stations.
This instrument calculates global solar radiation.	This instrument calculates direct solar radiation.

Solar Energy

Solar energy is the radiation from the Sun capable of producing heat, causing chemical reactions, or generating electricity. The total amount of solar energy incident on Earth is vastly in excess of the world's current and anticipated energy requirements. If suitably harnessed, this highly diffused source has the potential to satisfy all future energy needs. In the 21st century solar energy is expected to become increasingly attractive as a renewable energy source because of its inexhaustible supply and its nonpolluting character, in stark contrast to the finite fossil fuels coal, petroleum, and natural gas.

Solar panels: Solar panel array on a rooftop.

The Sun is an extremely powerful energy source, and sunlight is by far the largest source of energy received by Earth, but its intensity at Earth's surface is actually quite low. This is essentially because of the enormous radial spreading of radiation from the distant Sun. A relatively minor additional loss is due to Earth's atmosphere and clouds, which absorb or scatter as much as 54 percent of the incoming sunlight. The sunlight that reaches the ground consists of nearly 50 percent visible light, 45 percent infrared radiation, and smaller amounts of ultraviolet and other forms of electromagnetic radiation.

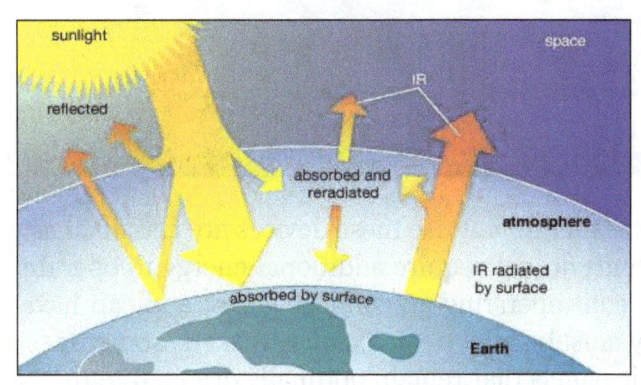

Solar energy: Reflection and absorption of solar energy. Although some incoming sunlight is reflected by Earth's atmosphere and surface, most is absorbed by the surface, which is warmed.

The potential for solar energy is enormous, since about 200,000 times the world's total daily electric-generating capacity is received by Earth every day in the form of solar energy. Unfortunately, though solar energy itself is free, the high cost of its collection, conversion, and storage still limits its exploitation in many places. Solar radiation can be converted either into thermal energy (heat) or into electrical energy, though the former is easier to accomplish.

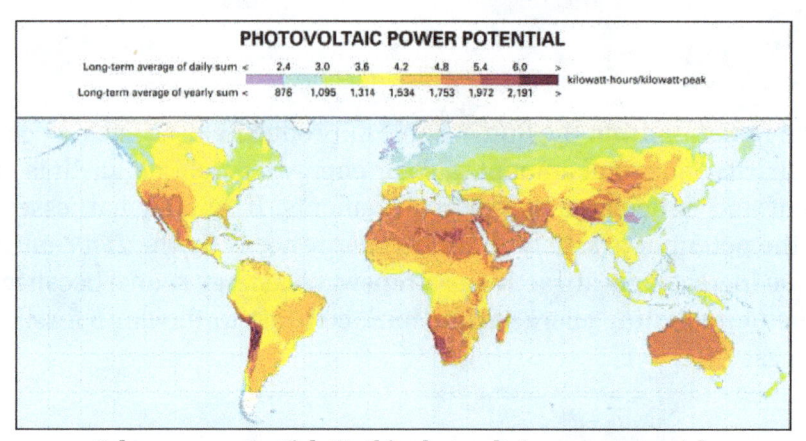

Solar energy potential: Earth's photovoltaic power potential.

Conversion of Solar Energy

Active Solar Energy

Active solar energy classifies technologies related to the use of solar energy that use mechanical or electrical equipment to improve performance or to process the energy obtained by converting it into electrical or mechanical energy. These equipments can be fans, water pumps, etc.

By contrast, solar systems that do not use these devices are classified as passive solar energy systems. Passive solar systems do not require additional energy to operate and therefore have zero operating costs, do not emit operating greenhouse gases, and can have low maintenance costs; passive solar technology must be fully treated. Solar trackers, sometimes used to improve the performance of photovoltaic panels that remain optimally oriented to the Sun, can be designed with some active or passive solar technology.

Types of Active Solar Energy

The applications of active solar energy can be classified into two types:

- Thermal Solar Energy,
- Photovoltaic Solar Energy.

Most low-temperature solar thermal collectors are usually located on fixed supports, but would have superior performance if they could follow the Sun on its way.

Thermal Solar Energy

Thermal solar energy takes advantage of the properties of thermodynamics to increase the temperature of a fluid by increasing its heat energy and, therefore, its entropy. In this form of use of solar energy, electricity is not generated except in thermoelectric solar plants.

The systems of hot water that are not based on the system used thermosyphon water pumps and fans to circulate the water within their circuits. For this reason, these systems would be classified as active solar energy. In contrast, solar water systems that use thermosiphon, the circulation of the water is carried out by means of the difference in densities between hot and cold water, they do not need external mechanisms and, therefore, they are classified as passive solar energy systems.

The majority of solar collectors are placed on fixed supports with a calculated orientation so that they can capture as much solar radiation during the day and throughout the year. But these collectors would be more efficient if they could modify their orientation towards the Sun depending on the day of the year and the time of day. There are systems to be able to modify the orientation of the solar collectors. In case these systems are mechanical, they would be active solar energy systems.

In the case of concentrated thermal solar energy, it is necessary to install solar monitoring systems. This type of installation needs to focus the solar radiation received at a point; therefore the mirrors have to adapt their function in relation to the position of the Sun. Generally, the function of this type of system is used in thermoelectric solar plants with the aim of generating electricity.

Photovoltaic Solar Energy

Photovoltaic solar energy is clearly an active solar energy system. Thanks to the photovoltaic effect, photovoltaic panels can generate electricity that will later go through transformers and other external elements. Similar to solar thermal energy, solar panels can be oriented by means of a small electric motor to orient them more efficiently towards the Sun.

Solar Tracker

Solar tracker is a system that positions an object at an angle relative to the Sun. The most-common applications for solar trackers are positioning photovoltaic (PV) panels (solar panels) so that they remain perpendicular to the Sun's rays and positioning space telescopes so that they can determine the Sun's direction. PV solar trackers adjust the direction that a solar panel is facing according to the position of the Sun in the sky. By keeping the panel perpendicular to the Sun, more sunlight strikes the solar panel, less light is reflected, and more energy is absorbed. That energy can be converted into power.

Solar panel: Solar panels in a field in La Calahorra, Granada, Spain.

Solar tracking uses complex instruments to determine the location of the Sun relative to the object being aligned. These instruments typically include computers, which can process complicated algorithms that enable the system to track the Sun, and sensors, which provide information to a computer about the Sun's location or, when attached to a solar panel with a simple circuit board, can track the Sun without the need for a computer.

Solar tracker; solar panel: A solar tracker adjusting the direction of a solar panel, keeping the panel perpendicular to the Sun in order to maximize the amount of sunlight that strikes the panel.

Studies have shown that the angle of light affects a solar panel's power output. A solar panel that is exactly perpendicular to the Sun produces more power than a solar panel that is not perpendicular. Small angles from perpendicular have a smaller effect on power output than larger angles. In addition, Sun angle changes north to south seasonally and east to west daily. As a result, although tracking east to west is important, north to south tracking has a less-significant impact.

Solar trackers provide significant advantages for renewable energy. With solar tracking, power output can be increased by about 30 to 40 percent. The increase in power output promises to open new markets for solar power. However, solar trackers have several important disadvantages. A static solar panel may have a warranty that spans decades and may require little to no maintenance. Solar trackers, on the other hand, have much shorter warranties and require one or more actuators to move the panel. These moving parts increase installation costs and reduce reliability; active tracking systems may also use a small amount of energy (passive systems do not require additional energy). Computer-based algorithm solar trackers are more expensive, require additional maintenance, and become obsolete much faster than static solar panels, since they use fast-evolving electronic components with parts that may be difficult to replace in relatively short periods of time.

Solar Cell

A solar cell (also known as a photovoltaic cell or PV cell) is defined as an electrical device that converts light energy into electrical energy through the photovoltaic effect. A solar cell is basically a p-n junction diode. Solar cells are a form of photoelectric cell, defined as a device whose electrical characteristics – such as current, voltage, or resistance – vary when exposed to light.

Individual solar cells can be combined to form modules commonly known as solar panels. The common single junction silicon solar cell can produce a maximum open-circuit voltage of approximately 0.5 to 0.6 volts. By itself this isn't much – but remembers these solar cells are tiny. When combined into a large solar panel, considerable amounts of renewable energy can be generated.

Construction of Solar Cell

A solar cell is basically a junction diode, although its construction it is little bit different from conventional p-n junction diodes. A very thin layer of p-type semiconductor is grown on a relatively thicker n-type semiconductor. We then apply a few finer electrodes on the top of the p-type semiconductor layer.

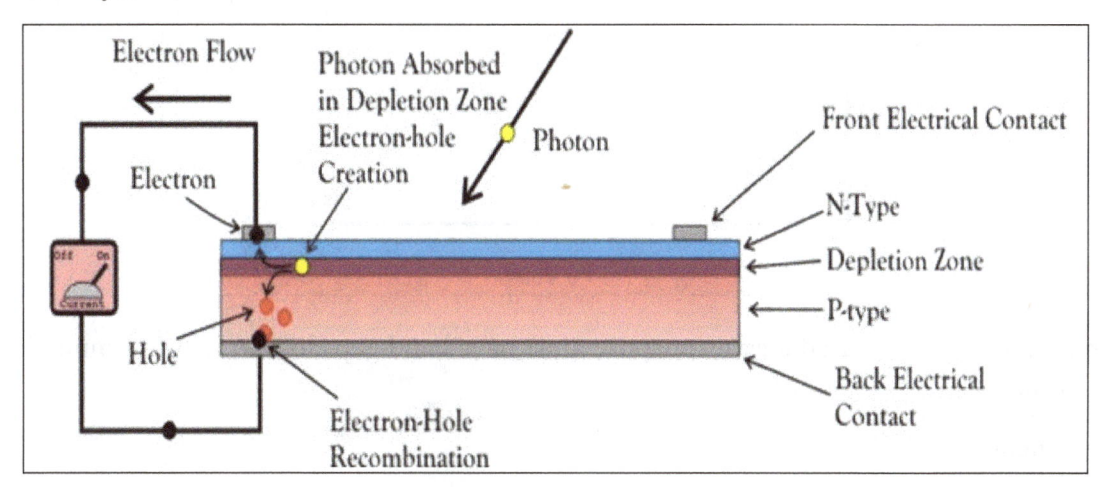

These electrodes do not obstruct light to reach the thin p-type layer. Just below the p-type layer there is a p-n junction. We also provide a current collecting electrode at the bottom of the n-type layer. We encapsulate the entire assembly by thin glass to protect the solar cell from any mechanical shock.

Working Principle of Solar Cell

When light reaches the p-n junction, the light photons can easily enter in the junction, through very thin p-type layer. The light energy, in the form of photons, supplies sufficient energy to the junction to create a number of electron-hole pairs. The incident light breaks the thermal equilibrium condition of the junction. The free electrons in the depletion region can quickly come to the n-type side of the junction.

Similarly, the holes in the depletion can quickly come to the p-type side of the junction. Once, the newly created free electrons come to the n-type side, cannot further cross the junction because of barrier potential of the junction.

Similarly, the newly created holes once come to the p-type side cannot further cross the junction became of same barrier potential of the junction. As the concentration of electrons becomes higher in one side, i.e. n-type side of the junction and concentration of holes becomes more in another side, i.e. the p-type side of the junction, the p-n junction will behave like a small battery cell. A voltage is set up which is known as photo voltage. If we connect a small load across the junction, there will be a tiny current flowing through it.

V-I Characteristics of a Photovoltaic Cell

Materials used in Solar Cell

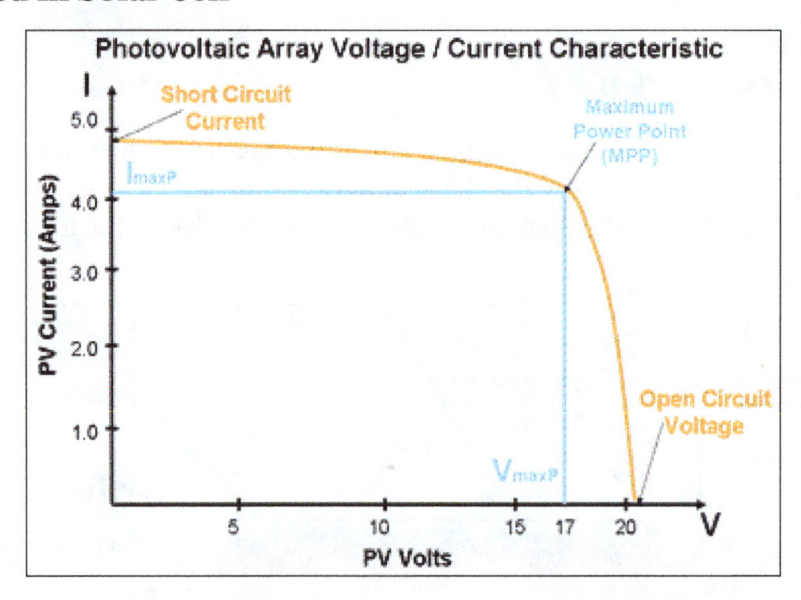

The materials which are used for this purpose must have band gap close to 1.5ev. Commonly used materials are:

- Silicon
- GaAs
- CdTe
- CuInSe2

Criteria for Materials to be used in Solar Cell

- Must have band gap from 1ev to 1.8ev.
- It must have high optical absorption.
- It must have high electrical conductivity.
- The raw material must be available in abundance and the cost of the material must be low.

Advantages of Solar Cell

- No pollution associated with it.
- It must last for a long time.
- No maintenance cost.

Disadvantages of Solar Cell

- It has high cost of installation.
- It has low efficiency.
- During cloudy day, the energy cannot be produced and also at night we will not get solar energy.

Uses of Solar Generation Systems

- It may be used to charge batteries.
- Used in light meters.
- It is used to power calculators and wrist watches.
- It can be used in spacecraft to provide electrical energy.

Solar Cell Efficiency

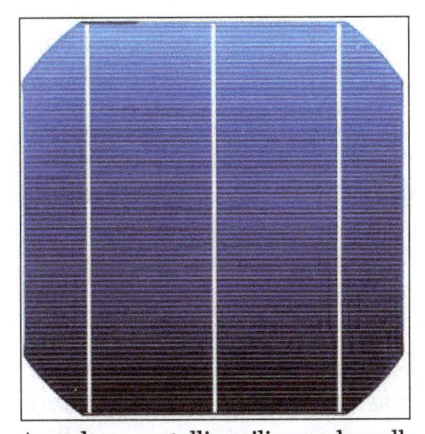

A modern crystalline silicon solar cell.

Efficiency is the comparison of energy output to energy input of a given system. For solar photovoltaic (PV) cells, this means the ratio of useful electrical energy they produce to the amount of solar energy incident on the cell under standardized testing conditions. Although some experimental solar cells have achieved efficiencies of close to 50%, most commercial cells are below 30%. Unlike the car not efficiency which limits the thermal efficiency of heat engines, the efficiency of solar cells is limited by something called the "band gap energy".

Band Gap Energy

The physics of semiconductors requires a minimum amount of energy to remove an electron from a crystal structure, known as the band gap energy. For solar cells, this energy is provided by particles of light called photons, which are tiny packets of electromagnetic radiation released from the Sun. Sunlight contains a wide spectrum of photons with different wavelengths and energy. If a photon strikes a solar cell and has less energy than the band gap, it gets absorbed as thermal energy. If it has sufficient energy it has a chance of "knocking" an electron loose and producing an electrical current. For silicon, the band gap energy is 1.12 electron volts. The longest wavelength (which corresponds to the lowest energy) that is capable of removing an electron is 1.1 µm. From figure it can be seen that the visible light and ultraviolet light have sufficient energy, but a good portion of infrared light cannot be used. This means that ~1/4 of the light from the Sun is not sufficient enough to create electricity.

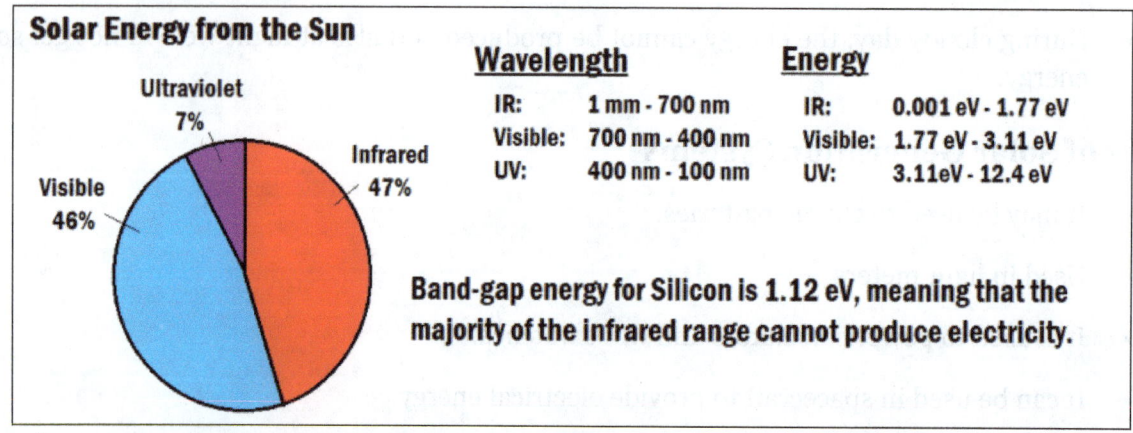

Solar Energy from the Sun

- Ultraviolet 7%
- Visible 46%
- Infrared 47%

Wavelength
- IR: 1 mm - 700 nm
- Visible: 700 nm - 400 nm
- UV: 400 nm - 100 nm

Energy
- IR: 0.001 eV - 1.77 eV
- Visible: 1.77 eV - 3.11 eV
- UV: 3.11 eV - 12.4 eV

Band-gap energy for Silicon is 1.12 eV, meaning that the majority of the infrared range cannot produce electricity.

The solar energy received by the Earth, and the corresponding energies of its photons. Efficiency of a PV cell is largely effected by the amount of incoming light that can cause current to flow.

Even from the light that can be absorbed, there is still a problem. Any energy above the band-gap energy will be transformed into heat. This also cuts the efficiency because that heat energy is not being used for any useful task. Of the electrons that are made available, not all of them will actually make it to the metal contact and generate electricity. This is because some of them will not be accelerated sufficiently by the voltage inside the semiconductor. Because of the reasons listed, the theoretical efficiency of silicon PV cells is about 33%.

Increasing Efficiency

There are ways to improve the efficiency of PV cells, all of which come with an increased cost. One way is to decrease the number of semiconductor impurities and crystal structure deformations. This can be achieved through the production of monocrystalline, or "single-crystal" cells. A more pure and uniform cell has a higher chance of interacting with incoming photons. Another method is to use a more efficient semiconducting material such as Gallium Arsenide. Although it's much more rare and expensive than silicon, gallium arsenide has an optimal band-gap of 1.4 electron volts, allowing for a higher percentage of the Sun's energy to be harnessed. Multiple layers of semiconductor material called p-n junctions can also be used to increase cell efficiency. These multi-junction cells harness energy from multiple sections of the solar spectrum as each junction has a different band gap energy. Efficiency can also be increased through concentrated photovoltaics. This method involves concentrating the Sun's energy through various methods to increase the intensity of energy hitting the solar cell.

Power Degradation

Efficiency of solar cells and solar panels are known to decrease over time, outputting less energy every year. This is due to a variety of factors including UV exposure and weather cycles.

Solar Inverter

A solar inverter can be defined as an electrical converter that changes the uneven DC (direct current) output of a solar panel into an AC (alternating current). This current can be used for different applications like in a viable electrical grid otherwise off-grid electrical network. In a PV system, it is a dangerous BOS (balance of system) component that allows the utilization of normal AC powered apparatus. These inverters have some functions with PV arrays like tracking of utmost PowerPoint & protection of anti-islanding. If we are using a solar system for a home, the selection & installation of the inverter is important. So, an inverter is an essential device in the solar power system.

solar-inverter.

Solar Inverter and it's Working

The working principle of the inverter is to use the power from a DC Source such as the solar panel and convert it into AC power. The generated power range will be from 250 V to 600 V. This conversion process can be done with the help of a set of IGBTs (Insulated Gate Bipolar Transistors). When these solid-state devices are connected in the form of H-Bridge, then it oscillates from the DC power to AC power.

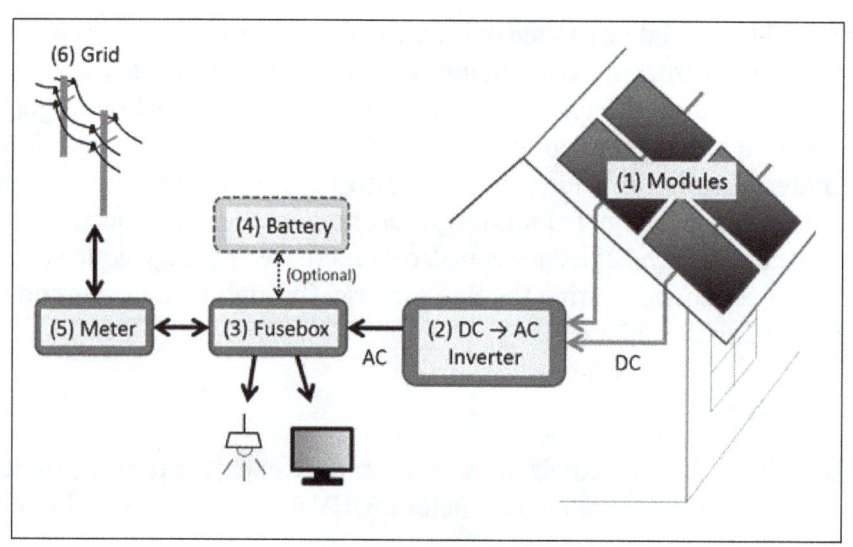

Solar-inverter-working.

A step-up transformer is employed so that the AC power can be obtained & can be fed to the grid. A few designers have started designing inverters without transformer which have high efficiency as compared with the inverters which have a transformer.

In any solar inverter system, a pre-programmed microcontroller is used to execute different algorithms exactly. This controller increases the output power from the solar panel with the help of the MPPT (Maximum Power Point Tracking) algorithm.

Types of Solar Inverters

Types-of-solar-inverters.

The classification of solar inverters can be done based on the application which includes the following:

String Inverter

This kind of solar panel is arranged in the form of a string and many strings are allied to a single string inverter. Every string holds the DC power where it is altered into AC power used like electricity. Based on the installation size, you may have many string inverters where each string gets DC power from some strings. These inverters are good for installations where the panels are arranged on a single plane to avoid facing in different directions.

String inverters can also be used with power optimizers as they are module-level power electronics that are mounted at the module level, consequently, every solar panel has one. Manufacturers of the solar panels use power optimizers with their devices & sell as one solution called a smart module so that installation can be made easier. Power optimizers give many benefits like microinverters, but they are less expensive. So it can be a good choice among using inverters like strictly string otherwise micro inverters.

Central Inverters

These are related to string inverters however they are larger & support additional strings of solar panels. Rather than running strings openly to the inverter, the strings are allied together in a general combiner box so that the DC power runs toward the middle inverter wherever it is transformed to AC power. These inverters needless connections of components, however, they need a pad as well as combiner box as they are suitable for huge installations through reliable production across the array.

The range of these inverters is from MWs to the hundreds of KWs and they handle up to 500kW for each area. These are not used in homes but used generally for huge commercial installations & utility-scale solar farms.

Microinverters

These inverters are a good choice for commercial as well as residential purposes. Same as power optimizers, these are also module-level electronics because one inverter is mounted on every panel. Microinverters alter power from DC to AC exact at the panel, so they don't need a string type inverter.

Also, due to the conversion of panel-level, if the performance of panels is shaded then the residual panels won't be exposed. These inverters monitor the function of every single panel, whereas string inverters illustrate the act of every string to make the inverters good at installation By using these inverters there are many benefits as they optimize every solar panel independently. It transmits more energy particularly if you have an incomplete shade situation.

Battery based Inverter

The growth in battery-based inverters is increased day by day. These are uni-directional and include both an inverter & battery charger. The operation of this can be done with the help of a battery. These inverters are separate grid-tied, grid-interactive and off-grid, based on the UL design & rating. The main benefit of this is, they give nonstop operation for critical loads based on the grid condition. In all occurrences, these inverters handle power between the grid & the array while charging the batteries, and they monitor the status of battery & controls how they are charged.

Hybrid Inverter

This inverter is also known as a multi-mode inverter and allows plugging batteries into the solar power system. It interfaces the battery through a method known as DC coupling. Electronics manage the charging & discharging of the battery. So there is a quite incomplete choice on these inverters.

Advantages of Solar Inverter

The main benefits of solar inverter include the following:

- Solar energy decreases the greenhouse effect as well as abnormal weather change.
- By using solar products, we can save money by reducing electricity bills.
- The solar inverter is used to change DC to AC and this is a reliable source of energy.
- These inverters empower small businesses by reducing their energy needs & requirements.
- These are multifunctional devices as they preprogrammed to alter DC to AC which assists large energy consumers.
- Easy to set up & more reasonable compared with generators.
- Maintenance is easy as they work well even with usual maintenance.

Disadvantages of Solar Inverter

The main drawbacks of solar inverter include the following:

- This kind of inverters is expensive to afford.
- Sunlight is necessary to generate sufficient electricity.
- It requires a huge space for installation.
- It requires a battery to work at night time to provide proper electricity to the home, commercial, etc.

Solar Constant

Solar constant is the total radiation energy received from the Sun per unit of time per unit of area on a theoretical surface perpendicular to the Sun's rays and at Earth's mean distance from the Sun. It is most accurately measured from satellites where atmospheric effects are absent. The value of the constant is approximately 1.366 kilowatts per square metre. The "constant" is fairly constant, increasing by only 0.2 percent at the peak of each 11-year solar cycle. Sunspots block out the light and reduce the emission by a few tenths of a percent, but bright spots, called plages, that are associated with solar activity are more extensive and longer lived, so their brightness compensates for the darkness of the sunspots. Moreover, as the Sun burns up its hydrogen, the solar constant increases by about 10 percent every billion years.

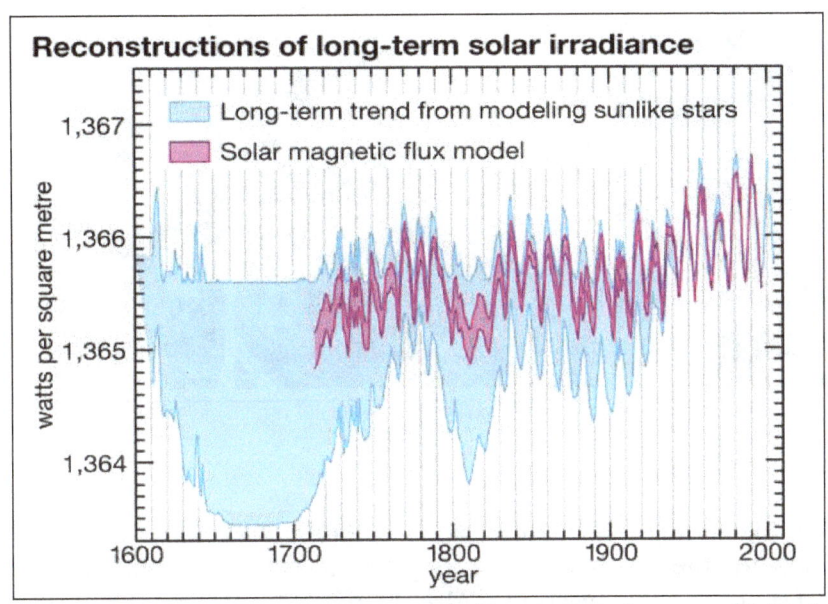

Reconstructions of long-term solar irradiance

Changes in the solar constant from 1600 to 2000. The blue region is from a model that is based on observations of stars such as the Sun, and the purple region is based on the effect of the solar magnetic flux on bright regions called faculae.

Advantages of Solar Energy

Renewable Energy Source

Among all the benefits of solar panels, the most important thing is that solar energy is a truly renewable energy source. It can be harnessed in all areas of the world and is available every day. We cannot run out of solar energy, unlike some of the other sources of energy.

Solar energy will be accessible as long as we have the sun, therefore sunlight will be available to us for at least 5 billion years when according to scientists the sun is going to die.

Reduces Electricity Bills

Since you will be meeting some of your energy needs with the electricity your solar system has generated, your energy bills will drop. How much you save on your bill will be dependent on the size of the solar system and your electricity or heat usage.

For example, if you are a business using commercial solar panels this switch can have huge benefits because the large system size can cover large chunks of your energy bills.

Diverse Applications

Solar energy can be used for diverse purposes. You can generate electricity (photovoltaics) or heat (solar thermal). Solar energy can be used to produce electricity in areas without access to the energy grid, to distil water in regions with limited clean water supplies and to power satellites in space.

Solar energy can also be integrated into the materials used for buildings. Not long ago Sharp introduced transparent solar energy windows.

Low Maintenance Costs

Solar energy systems generally don't require a lot of maintenance. You only need to keep them relatively clean, so cleaning them a couple of times per year will do the job. Most reliable solar panel manufacturers' offer 20-25 years warranty. Also, as there are no moving parts, there is no wear and tear. The inverter is usually the only part that needs to be changed after 5-10 years because it is continuously working to convert solar energy into electricity and heat (solar PV vs. solar thermal). Apart from the inverter, the cables also need maintenance to ensure your solar power system runs at maximum efficiency.

So, after covering the initial cost of the solar system, you can expect very little spending on maintenance and repair work.

Technology Development

Technology in the solar power industry is constantly advancing and improvements will intensify in the future. Innovations in quantum physics and nanotechnology can potentially increase the effectiveness of solar panels and double, or even triple, the electrical input of the solar power systems.

Disadvantages of Solar Energy

Cost

The initial cost of purchasing a solar system is fairly high. This includes paying for solar panels, inverter, batteries, wiring, and the installation. Nevertheless, solar technologies are constantly developing, so it is safe to assume that prices will go down in the future.

Weather-Dependent

Although solar energy can still be collected during cloudy and rainy days, the efficiency of the solar system drops. Solar panels are dependent on sunlight to effectively gather solar energy. Therefore, a few cloudy, rainy days can have a noticeable effect on the energy system. You should also take into account that solar energy cannot be collected during the night.

On the other hand, if you also require your water heating solution to work at night or during wintertime, thermodynamic panels are an alternative to consider.

Solar Energy Storage Is Expensive

Solar energy has to be used right away, or it can be stored in large batteries. These batteries, used in off-the-grid solar systems, can be charged during the day so that the energy is used at night. This is a good solution for using solar energy all day long but it is also quite expensive.

In most cases, it is smarter to just use solar energy during the day and take energy from the grid during the night (you can only do this if your system is connected to the grid). Luckily your energy demand is usually higher during the day so you can meet most of it with solar energy.

Uses a Lot of Space

The more electricity you want to produce, the more solar panels you will need, as you want to collect as much sunlight as possible. Solar PV panels require a lot of space and some roofs are not big enough to fit the number of solar panels that you would like to have.

An alternative is to install some of the panels in your yard but they need to have access to sunlight. If you don't have the space for all the panels that you wanted, you can opt for installing fewer to still satisfy some of your energy needs.

Associated with Pollution

Although pollution related to solar energy systems is far less compared to other sources of energy, solar energy can be associated with pollution. Transportation and installation of solar systems have been associated with the emission of greenhouse gases.

There are also some toxic materials and hazardous products used during the manufacturing process of solar photovoltaic systems, which can indirectly affect the environment. Nevertheless, solar energy pollutes far less than other alternative energy sources.

Solar Energy Storage System

A solar system operating in conjunction with batteries may offer the system owner far more functionality than a solar system operating without such additional power. A system with batteries can fully power an average home for several hours as long as that home is connected to the grid. In blackouts, batteries can power selected circuits in the home around the clock because they get recharged in the daytime by the solar panels.

A solar energy storage system consists of four main parts:

- Solar panels – Provide electricity to the system with sufficient sunlight.

- Solar charge controllers – Manages the power going into the batteries, and prevents reverse current which would drain the batteries when the sun isn't shining.

- Inverter – Converts DC power from the solar panels or the batteries into AC power for the home or grid.

- Batteries – Stores DC power from the solar panels for later use in the home.

During low insolation times, solar energy storage system enables delivery of more power than what is generated by the solar electric or thermal plant, and so it enables to match the generation of energy with the load demand.

Classification of Solar Energy Storage System

The solar energy storage systems can be classified as follows:

- The thermal energy storage system.
- Chemical energy storage system.
- Electrical energy storage system.
- Hydrogen energy storage system.
- Electromagnetic energy storage system.
- Biological storage system.

Thermal Energy Storage

Thermal acid batteries are the most commonly used means in chemical energy storage system. The advantages are (i) good working efficiency (up to 80%), (ii) low cost, (iii) rapid change from charging to discharging mode and (iv) slow discharge rate. A storage battery takes electrical energy generated by solar radiation and stores it as chemical energy. It later supplies electric energy by converting this stored energy.

Electrical Energy Storage

A capacitor is used to store electrical energy in electrostatic field when it is charged. The capacitor of large capacity is required to store a significant amount of energy.

Hydrogen Energy Storage

The electrical energy is used to decompose water by the electrolysis reaction into hydrogen and oxygen. These substances can be recombined to release the stored energy when required.

Electromagnetic Energy Storage

The electrical energy is used to store energy in a magnetic field. The resistance of the coil wire is made almost negligible so that the stored energy in the coil is not dissipated out and stored energy in the magnetic field can be maintained indefinitely. The electromagnetic energy storage requires the use of superconducting materials. These materials develop almost zero resistance to electricity flow when cooled below a critical or transition temperature. This method of storing electromagnetic energy is also called super conducting magnetic energy storage (SMES). The electric energy can be recovered when coil is discharged.

Biological Storage

The solar energy is stored in plant by a process known as photosynthesis. Photosynthesis is the process in which organic compounds are formed in green plant using carbon from atmospheric carbon dioxide in the presence of sunlight. The plants on decaying from biomass which can be converted into various types of solid, liquid and gaseous fuels.

Sensible Heat Storage

Thermal energy is stored in this type of storage by virtue of heat capacity and temperature difference developed during charging and discharging. The temperature of the storage material rises when thermal energy is absorbed and temperature drops when thermal energy is taken out. In this storage, the charging and discharging can be performed reversibly for an unlimited number of time. The sensible heat storage can be liquid media storage and solid media storage.

Sensible Heat Storage by Water

Water is considered as the most suitable media for storage below 100° C. liquid such as oils, liquid metals and molten salt are also used as liquid media storage. The water thermal energy storage can be short term and long term. A short-term thermal energy storage system has a well-insulated storage tank as shown in figure. The storage in such tank is economical for few days only as heat losses over long duration make the storage uneconomical.

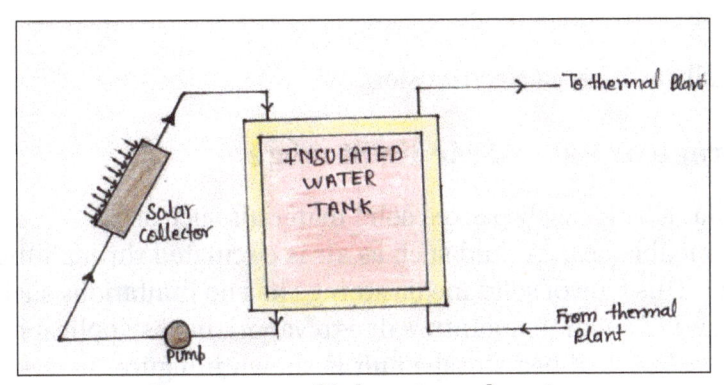

Short term sensible heat storage by water.

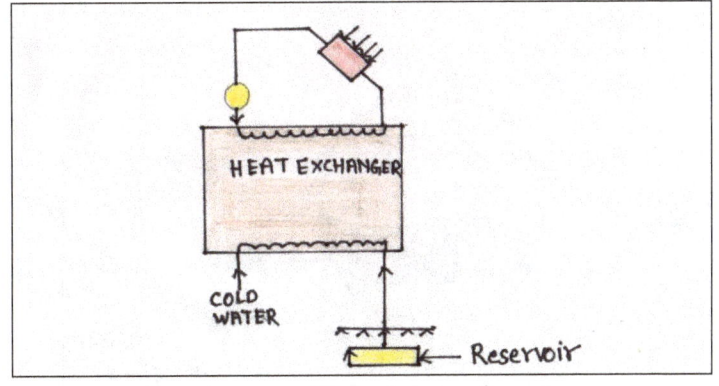

Long term sensible heat storage by water.

Long-term sensible heat storage by water is possible in underground reservoir having special insulation. In this system, water is heated in charging mode by passing it through a heat exchanger and then it is stored in an underground reservoir. In the discharge mode, the hot water is made to flow back through the heat exchanger, where release the stored energy as shown in figure but with reverse circulation.

The advantages of this storage system are:

- It is abundantly available.
- It is inexpensive.
- It has high specific heat which enables to store more heat per unit mass.
- It has low viscosity requiring less energy to pump through the pipe system.
- It can be used for both storage and working medium.
- It is stable.
- It has no harmful effect.

But water has the following disadvantages:

- It has limited temperature range of 0-100 °C.
- It results in the corrosion of pipes.
- It can leak easily as it has surface tension.

Solid Media Storage or Packed Media Storage

This type of storage has a bed loosely packed solid materials such as rocks, sand, concrete, pebbles and metals to store sensible heat. A fluid such as air is circulated through the bed to add remove heat from the storage. This type of solid media storage has no limitations such as (i) low temperature due to freezing and (ii) high temperature due to vaporizing as applicable in the case of liquid media storage. A typical packed bed storage unit is shown in figure. It consists of a container, a screen to support the bed inlet duct and outlet duct.

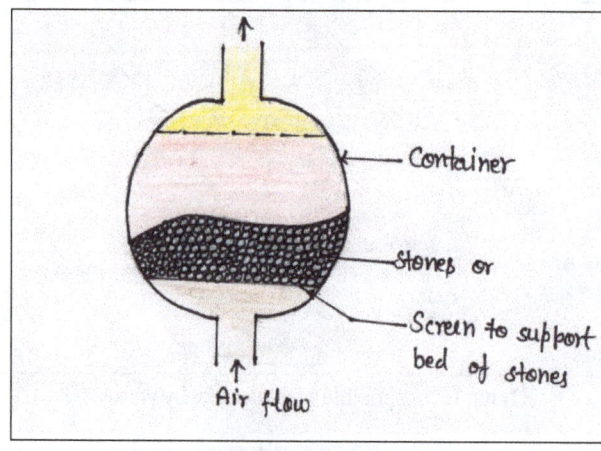

Solid media storage.

The charging or adding of heat is done by passing hot air through the bed in one direction and the removable of heat is done by passing the normal air through the bed in the opposite direction.

The advantages of solid media storage are as follows:

- Stones or pebbles are abundantly available,
- Low cost,
- Non-combustible,
- Easy to handle,
- Possibility of high storage temperature,
- No freezing point during heat removal,
- No corrosion problem,
- No requirement of heat exchanger.

The disadvantages are as follows:

- The size of the storage container should be large,
- Simultaneously charging and discharging of energy is impossible,
- Large pressure drop needs high capacity air blower.

Thermodynamic Considerations

The storage of sensible heat is based - thermodynamically speaking - on the increase of enthalpy of the material in the store, either a liquid or a solid in most cases. The sensible effect is a change in temperature. The thermal capacity - this is the heat which can be put in the store or withdrawn from it - can be obtained by the equation,

$$Q_{12} = m \int_{T_1}^{T_2} c_p(T) dT$$

With the specific heat capacity c_p being a function of temperature T, and m mass.

For a temperature independent c_p this becomes simply,

$$Q_{12} = m c_p (T_2 - T_1)$$

Where, the unit of Q_{12} is, e. g., J. The symbol m stands for the store mass and T_2 denotes the material temperature at the end of the heat absorbing (charging) process and T_1 at the beginning of this process. This heat is released in the respective discharging process.

For pure solids (especially heavy elements), the specific heat per mole of a substance is approximately

3R (Dulong-Petite rule), with R being the molar gas constant (R=8.31441 kJ/kmolK). Thus, the molar thermal energy q_{mol} stored in solids can be approximated by,

$$q_{mol} \approx 3R \otimes T$$

And q_{mol} is measured e. g. in kJ/kmol. Thus, approximately 25 kJ/kmol can be stored with a temperature difference of $\otimes T$ = 1K. With the molar mass M (kg/kmol), the thermal energy q stored per mass (store capacity in kJ/kg) is obtained,

$$q = c_p \otimes T = q_{mol}/M = 3R \otimes T/M$$

Heat has a quality, namely its temperature. This determines how much of the heat can be usefully applied according to the second law. This available energy is called Exergy Ex and can be obtained from,

$$E_x = Q_{12} - \frac{T_{am}}{T} Q_{12}$$

When the Kelvin temperature of the store T = constant during delivery of the heat, with Tam being the ambient temperature, e.g. 293 K.

For varying temperatures of the store (i.e. stores of a finite size) the exergy change is expressed by,

$$\Delta E_{x12} = (H_2 - H_1) - T_{am}(S_2 - S_1) - (p - p_{am})(V_2 - V_1)$$

With H being enthalpy, S entropy, V volume and p being pressure.

Under atmospheric pressure $p = p_{am}$ and with the 1st and 2nd law this reduces to,

$$\Delta E_{x12} = mc_p \left[(T_2 - T_1) - T_{am} \ln \frac{T_2}{T_1} \right]$$

From equation (previous) it can be observed that exergy changes are not linear with temperature as energy changes.

Exergy is low at low temperatures but increases steeply with an increase in temperature while the energy increase remains the same at low or high temperatures.

A second law evaluation of stores (i.e. by exergy) is more useful than a first law evaluation (by energy). This becomes obvious in the following example on the importance of temperature stratification in warm water stores:

Let us assume a container which is half filled with water of 50 °C (upper half) and 20 °C (lower half). If the two water layers will be fully mixed, the mean temperature is 35 °C. The energy in this container remains the same, provided no heat losses occur to the surroundings. Following now equation (4), assuming an ambient temperature of T = 293 °K, a loss in exergy can be calculated. It amounts to $\Delta E_x = 1.23 mc_p$. The energetic consideration does not show any change by the destruction of a thermal layering in the store; the exergetic consideration, however, does indicate the quality of the stored water which, with respect to temperature, has certainly decreased. Half of the container water with 50 °C is of more use - e. g. for taking a shower - than all of the water with 35 °C.

Solid Storage Materials

Properties of some solid materials are given in table.

Table: Thermophysical properties of some solid materials.

Material	M	θ	ρ	c	k	$10^{6}a$	$10^{-3}b$
	kg/kmolg	°C	kg/m³	kJ/kgK	W/mK	m²/s	J/m²ks$^{1/2}$
Aluminium 99.99 %	27	20	2700	0.945	238.4	93.3	24.66
Copper (commercial)	63.5	20	8300	0.419	372	107.	35.97
Iron	56	20	7850	0.465	59.3	16.3	14.7
Lead	207	20	11340	0.131	35.25	23.6	7.24
Brick (dry)		20	1800	0.840	0.5	0.33	0.87
Concrete (gravelly)		20	2200	0.879	1.279	0.66	1.57
Granite		20	2750	0.892	2.9	1.18	2.67
Graphite (solid)	12	20	2200	0.609	155	120	14.41
Limestone		20	2500	0.741	2.2	1.19	2.02
Sandstone		20	2200	0.710	1.8	1.15	1.68
Slag		20	2700	0.836	0.57	0.25	1.13
Soil (clay)		20	1450	0.880	1.28	1.00	1.28
Soil (gravelly)		20	2040	1.840	0.59	0.16	1.49

Metals and graphite have high values of temperature diffusivity α and heat diffusivity b. Thus, these materials are best suited for quick charging and discharging. Other materials such as bricks and stones are less advantageous in this respect, it takes longer for them to heat and cool but in volume storage capacity (ρ · c) granite can compete with aluminium.

Solid materials can be heated to very high temperatures (e. g. magnesia bricks in Cowper regenerators to 1.000 °C). They are usually chemically inert and, in the form of rocks and pebbles, abundant and cheap.

Pebble bed and rock pile stores consist of loosely packed material in a container. Heat has to be carried to or from such stores by a heat carrier such as air or water. These beds provide a large surface area for heat transfer in their dispersed materials; the internal heat losses are only small because the particles have little surface contact to surrounding particles and thus, little conduction losses.

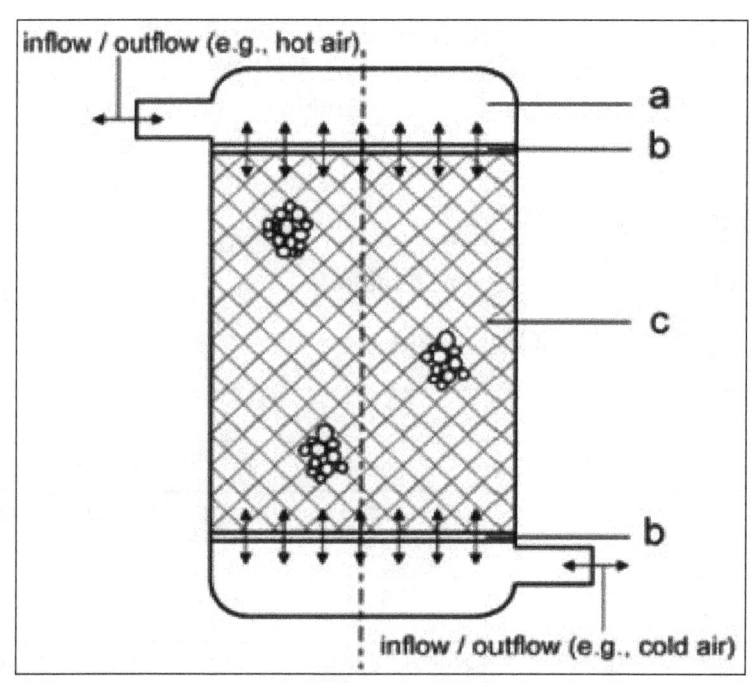

Container with solid particles for thermal storage: (a) air distribution chamber, (b) grating, (c) particles.

Reduced insulation around the store container is sufficient when air - with a small thermal conductivity - is used as a heat carrier and fills the gaps between the particles. Such particle beds, however, cause a large pressure drop in the fluid flow. This has to be compensated by pumps or compressors. Another drawback is the large storage volume required. Conveniently, this is provided in underground installations.

In ground stores, heat is stored in soil, stones or solid rocks, especially for low temperature storage. Cast iron is an inexpensive material with a high volume storage capacity ($\rho \cdot c$). It is used for high temperature storage together with oil as the heat carrier. As solid materials for storage always have to be combined with fluid heat carriers, properties of the solids have to be taken into account which affects fluid flow and heat transfer, e. g., particle and container size, mechanical durability against abrasion and thermal cycling, hygroscopy, etc. In figure, a storage bin for solid materials with gas as a heat carrier is shown. If the packing density is high, e.g. for small solid particles, the store capacity is high, but the pressure drop is also high. A proper particle size which should be rather uniform has to be found.

The thermal stratification is well maintained in a solid-material heat store because natural convection in the heat carrier is suppressed by the granulate, and internal heat conduction is low. Containers as shown in figure are used - in form of pillars - as architectural components in buildings. The combination of hollow bricks and air has been known as hypocaust-heating systems since antiquity.

Latent Heat Storage

In latent heat storage, heat energy is stored by virtue of latent heat which is required to bring about phase change of storage medium. The heat required to bring about phase change of a material is much larger compared to sensible heat change of the same material. The phase change of material also involves absorption or release of large quantity of heat energy at constant temperature, which is impossible in the case of sensible heating and cooling. Therefore, latent heat storage system is more compact for a certain heat storage compared to sensible heat storage system. The phase changes which can be used for storage system are solid-solid, solid- gas, solid – liquid and liquid-gas. Solid – gas and liquid –gas transformation involves large volume changes, thereby making such storage system impractical and complex. However, solid- solid transition involves transformation of the material from one crystalline form to another, thereby resulting in the transformation with small volume changes. Hence, such storage systems are practical and preferred in spite of small changes in latent heat possible during transformation.

Materials used for latent heat storage are called phase change materials (PCM). The amount of heat stored is calculated following equation,

$$Q = m \cdot \Delta h$$

Where, Q is the amount of heat stored in the material (J), m is the mas of storage material (kg), and Δh is the phase change enthalpy (J/kg). The best known and used PCM is water, used as ice for cold storage since early times.

For the phase-change storage media, salt hydrates called glauber's salt ($Na_2SO_4 \cdot 10H_2O$) are preferred. These have solid – liquid transformation. Besides hydrates, paraffins ($C_{18}H_{36}$) and non-paraffins (ester, fatty acid, alcohols and glycols) are also suitable for such storage. The hydrate crystals have water of crystallization and these can be represented by $X(Y)n \cdot mH_2O$ (one atom of X, n atoms of Y and m molecules of water in one crystal). When hydrate crystals are heated to transition temperature, these crystals release their water of crystallization and anhydrous salt (hydrates without water) get dissolved in the released water.

Different PCMs are needed for different energy storage temperatures. The available PCMs include organic PCMs, inorganic PCMs, and eutectic PCMs. One of the most important groups of organic PCMs is paraffin wax. Take paraffin (n-docosane) with a melting temperature of 42–44 °C as an example: it has a latent heat of 194.6 kJ/kg and a density of 785 kg/m³. The energy density is 42.4 kWh/m³. Nonparaffin organic PCMs include the fatty acids and glycols. Inorganic PCMs include salt hydrates, salts, metals, and alloys. Eutectic PCMs are a minimum-melting mixture of several different PCMs.

The problems faced with the use of salt hydrate for latent heat storage are as follows:

- The released water of crystallization is insufficient to dissolve all the solid salt produced on heating. The anhydrous salt settles down at the bottom of the container. The recrystallization becomes impossible on removal of heat. The process becomes irreversible and performance degradation takes place.

- Mechanical means (vibration or stirring), the suspension media or thickening agents have to be used to make the system work in reversible manner without performance degradation. The problem can also be resolved by limiting vertical height of the container.

- Heat of fusion is small (251 kJ/kg).

References

- What-is-pyranometer-construction-types-applications: elprocus.com, Retrieved 12, June 2020
- 5-advantages-and-5-disadvantages-of-solar-energy: greenmatch.co.uk, Retrieved 28, April 2020
- How-solar-energy-storage-systems-work: freedomforever.com, Retrieved 30, July 2020
- Active-solar-power, what-is-solar-energy: solar-energy.technology, Retrieved 25, May 2020
- What-is-a-solar-inverter-and-how-it-works: elprocus.com, Retrieved 18, April 2020

Chapter 2

Solar Collector and Thermal Technologies

Any device that is used to collect solar radiations is called a solar collector. There are several different types of solar collectors available for use, each with its own working principle. The technology of producing, storing, controlling, transmitting and getting work done by heat energy is known as thermal technology. Both these topics are discussed in detail in this chapter.

Solar Collector

A solar collector is an object that is used to collect energy from the sun, which it does by absorbing solar radiation and converting it into heat or electricity. The material type and coating on a solar collector are used to maximize solar energy absorption.

There are several different types of solar collectors available for use. One common type of solar collector is the flat plate collector. This solar collector has an absorber plate that is placed in a box enclosure. The absorber plate is oftentimes coated to better absorb heat energy. The top of the box, facing the sun, has a transparent cover that allows solar energy to pass through it but limits the amount of heat energy lost.

Evacuated tube collectors are another popular type of solar collector. An evacuated tube collector has its solar collection material placed inside a glass tube that has little to virtually no air inside of it. There are liquid-filled pipes in the vacuum-sealed glass container that are heated by the solar collector; this allows the collector to delivery energy. There are several other types of solar collectors available as well.

Solar collectors usually transfer energy by heating a liquid, which can be used to accomplish several things. For a residential solar collector system, it can be used to heat water for bathing, cooking or cleaning. The heated water can be used in industrial applications as well. For energy companies, solar farms typically take the heated liquid and convert the heat energy of the liquid into electricity and distribute it to their customers.

Heat from sun's rays can be harnessed to provide heat to a variety of applications. But in general, sun's rays are too diffuse to be used directly in these applications. So solar concentrators are used to collect and concentrate sun's rays to heat up a working fluid to the required temperature. Therefore, a solar concentrating collector is defined as a solar collector that uses reflectors, lenses or other optical elements to redirect and concentrate solar radiation onto a receiver.

The solar heat can be used as hot water, air or steam that can be readily deployed for meeting numerous applications in different sectors such as industrial process heating, power generation on a large scale, community cooking and space cooling/ heating. Solar collectors are classified as low, medium or high temperature collectors. Low – temperature collectors are used for smaller non-intensive requirements. Medium-temperature collectors are used for heating water or air for industrial and commercial use. High temperature collectors concentrate sunlight using mirrors or lenses and are used for fulfilling heating requirements up to 400 °C/20 bar pressure in industries.

Table: Temperature range of solar thermal technologies.

	Temperature Range
Low temperature heat	< 150 °C
Medium temperature heat	150 – 400 °C
High temperature heat	> 400 °C

Solar thermal technologies use various collectors to generate heat. A collector is a device for capturing solar radiation. Solar radiation is energy in the form of electromagnetic radiation from the infrared (long) to the ultraviolet (short) wavelengths. Solar collectors are either non-concentrating or concentrating. In the non-concentrating type, the collector area (i.e., the area that intercepts the solar radiation) is the same as the absorber area (i.e., the area absorbing the radiation). In these types the whole solar panel absorbs light. Concentrating collectors have a larger interceptor than absorber.

Non concentrating solar thermal collectors are generally used for low and medium temperature requirements. Solar water heating is the perfect example of a non - concentrating type of solar thermal application.

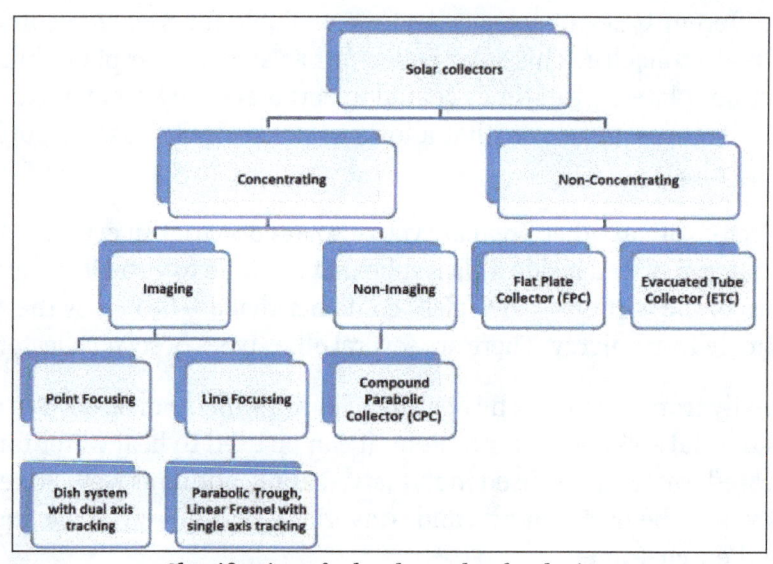

Classification of solar thermal technologies.

Table: Types of Solar Collectors.

Motion	Collector Type	Absorber Type	Concentration Ratio	Indicative Temperature Range (°C)
Stationary	Flat Plate Collector (FPC)	Flat	1	30-80
	Evacuated tube collector (ETC)	Flat	1	50-200
	Compound Parabolic Collector (CPC)	Tube	1-5	60-240
			5-15	60-300

Single Axis Tracking	Linear Fresnel reflector (LFR)	Tubular 1	10-40	60-250
	Parabolic Trough Collector (PTC)		15-45	60-300
Dual Axis Tracking	Parabolic dish reflector (PDR)	Point	100-1000	100-500
	Heliostat field collector (HFC)	Point	100-1500	150-2000

Non-Concentrating Technology

Non concentrating solar thermal collectors are generally used for low and medium energy requirements. Solar water heating is the perfect example of a non - concentrating type of solar thermal application. A solar water heater is a combination of an array of collectors, an energy transfer system, and a thermal storage system. In an active SWH (solar water heating) system, a pump is used to circulate the heat transferring fluid through the solar collectors. The amount of hot water produced from a solar water heater critically depends on design and climatic parameters such as solar radiation, ambient temperature, wind speed etc. Common collectors used for solar water heaters are - Flat Plate Collectors and Evacuated Tube Collectors.

Flat Plate Collectors

Flat Plate Collectors consist of a thin metal box with insulated sides and back, a glass or plastic cover (the glazing) and a dark colour absorber. The glazing allows most of the solar energy into the box whilst preventing the escape of much of the heat gained. The absorber plate is in the box painted with a selective dark colour coating, designed to maximize the amount of solar energy absorbed as heat. Running through the absorber plate are many fine tubes, through which water is pumped. As the water travels through these tubes, it absorbs the heat. This heated water is then gathered in a larger collector pipe to be transported into the hot water system.

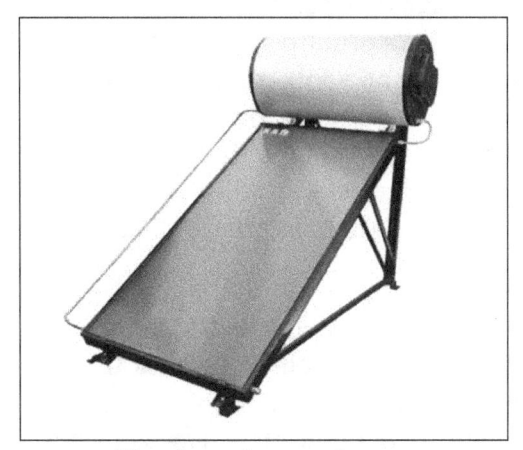

Flat plate solar water heater.

Evacuated Tube Collector

Evacuated tube collectors are more modern and more efficient in design. These can heat water to much higher temperatures and require less area. However, they are also correspondingly more expensive. Instead of an absorber plate, water is pumped though absorber tubes (metal tubes with a selective solar radiation absorbing coating), gaining heat before going into the collector pipe. Each absorber tube is housed inside a glass tube from which the air has been evacuated forming a vacuum. The glass tube allows solar radiation through to the absorber tube where it can be turned into heat. The vacuum eliminates convective as well as conductive heat loss and virtually all heat absorbed is transferred to the water.

Evacuated tube solar water heater.

Brief on Concentrating Solar Technologies (CSTs)

These systems utilise solar radiation to generate heat – as hot water, air or steam that can be readily deployed for meeting numerous applications in different sectors such as industrial process heating, power generation on a large scale and space heating/cooling. These applications make use of solar energy collectors as heat exchangers that transform solar radiation energy to internal energy of the transport medium (or heat transfer fluid, usually air, water, or oil). The solar energy thus collected is carried from the circulating fluid either directly to the hot water or space conditioning equipment or to a thermal energy storage tank from which can it be drawn for use at night and/or cloudy days.

Unlike the non - concentrating solar thermal systems, concentrating solar thermal systems use mirrors and lenses to reach higher temperature mainly for various industrial processes. Under the concentrating type - there are imaging and non - imaging technologies. Imaging technologies have a smaller range of acceptance angle compared to that of non - imaging technologies. For example, imaging concentration gives about 1/3 of the theoretical maximum for the design acceptance angle, that is, for the same overall tolerances for the system. Non - imaging optics, which have a larger acceptance angle range, can be used to approach the theoretical maximum.

Imaging technologies, which make use of mirrors and lenses, are divided in to two types each based on the focus and receiver type. Line Focus collectors track the sun along a single axis and focus the irradiance on a linear receiver. Point focus collectors track the sun along two axes and focus the irradiance at a single point receiver which allows for higher temperatures. Fixed receivers are stationary devices that remain independent of the plant's focusing device which eases the transport of collected heat. Mobile receivers move together with the focusing device which enables more energy to be collected. The different types of solar thermal technologies are depicted in the figure below.

Concentrating Solar Thermal (CST) Technology

Industrial heat is characterized by a wide diversity with respect to temperature levels, pressures and production processes to meet many different industrial process demands. Concentrated Solar Technologies (CSTs) track the sun's incoming radiation with mirror fields, which concentrate the energy towards absorbers, which then transfer it thermally to the working medium. The heated fluid or steam may reach high temperatures and may be used for various processes requiring heat.

CSTs can produce a range of temperatures, between 50 °C and up to over 400 °C, which can be used in a variety of industrial and commercial heat applications. The industries showing good potential for implementation of solar concentrators are food processing, dairy, paper and pulp, chemicals, textiles, fertilizer, breweries, electroplating, pharmaceutical, rubber, desalination and tobacco sectors. Any industrial/commercial establishments currently using steam/hot water for process applications can also employ CSTs with a minimum tinkering to the existing setup.

CST Technologies

The most mature concentrated solar technologies are:

- Point Focus Technology:
    - Fixed focus automatically tracked elliptical dish (Scheffler dish).
    - Fresnel Reflector based dish (ARUN dish).
    - Dual axis tracked paraboloid dish.
- B. Line Focus Technology:
    - Parabolic troughs collectors (PTC).
    - Linear Fresnel Reflector (LFR).
- C. Non-Focussing Technology:
    - Compound Parabolic Collectors(CPC).

Fixed Focus Elliptical Dish (Scheffler)

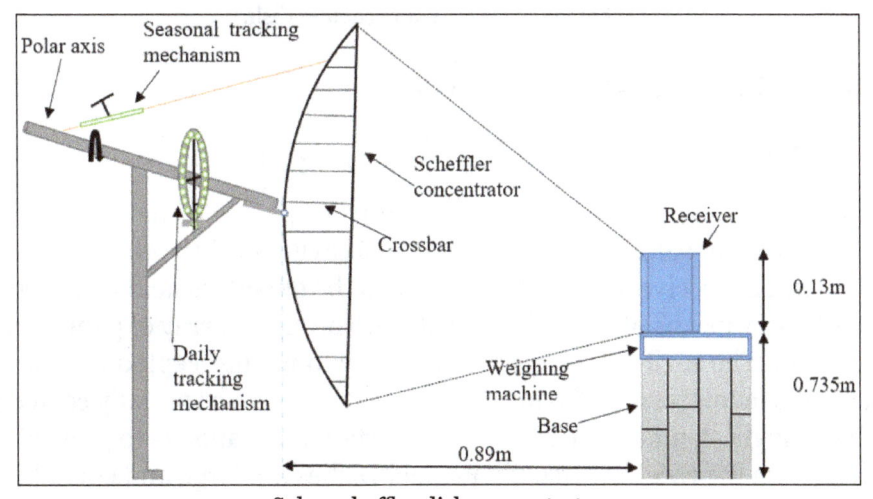

Solar scheffler dish concentrator.

Fixed Focus Elliptical Dish, often called a Scheffler concentrator, is a small lateral section of a paraboloid which concentrates sun's radiation over a fixed focus. The dish comprises of a large number of mirrors to reflect the sun rays to a fixed receiver which contains a working fluid to be heated. The Scheffler dish system works on the following principles:

- The parabolic reflective dish turns about north-south (N-S) axis parallel to earth's axis, tracking the sun's movement from morning (East) to evening (West).

- The parabolic reflector also performs change in inclination angle while staying directed to sun, in order to obtain sharp focal point. Adjustments for the seasonal variations in the sun's position (N - S direction) have to be made manually every few days by an operator.

- Focus always lies at the axis of rotation. It remains at a fixed position, where concentrated heat is captured and transferred to water through the receiver to generate hot water or high pressure steam.

- Water from header pipe passes to receiver (thermosyphon principle). At the receiver, the hot water or steam generated water and collected in the header pipe flows to the end use application.

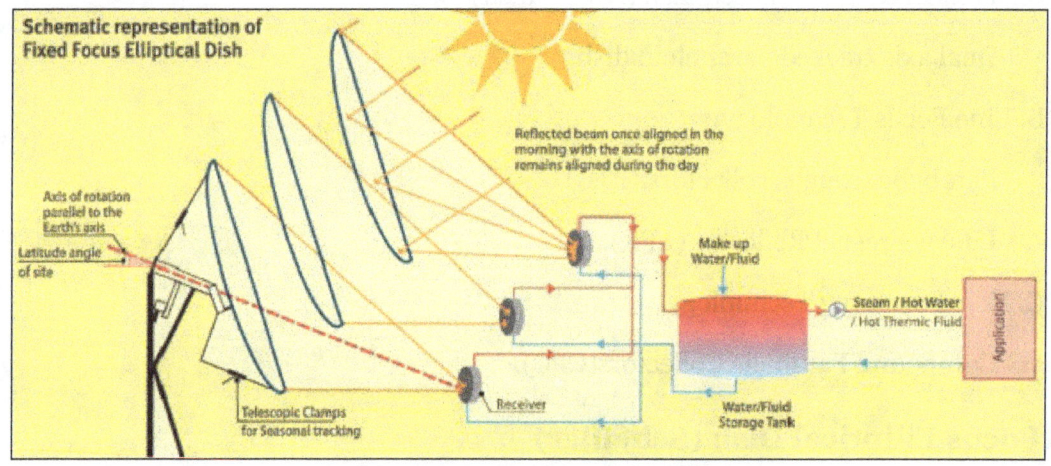

Schematics diagram for scheffler dish.

Fresnel Reflector based Dish (ARUN Dish)

Underlying Principle

Fresnel Reflector Based Dish (ARUN Dish) is made from panels of flat mirrors mounted on a frame such that the incident sunlight is reflected on to a cavity receiver which is specially designed to reduce heat losses. The receiver which is insulated on the outside is held in a fixed position in relation to the reflectors by means of a suitable structure. The entire array of panels with the receiver move to track the sun. The cavity receivers allow energy to be intercepted by a small aperture or opening which results in low losses. The inside of the cavity may be specially coated to increase its absorption of the sunlight that falls on it. There are certain mechanisms to prevent or reduce convective heat losses from the receiver. Usually Fresnel Dishes are large and could have an aperture area of 100 m² or 160 m².

Solar Collector and Thermal Technologies

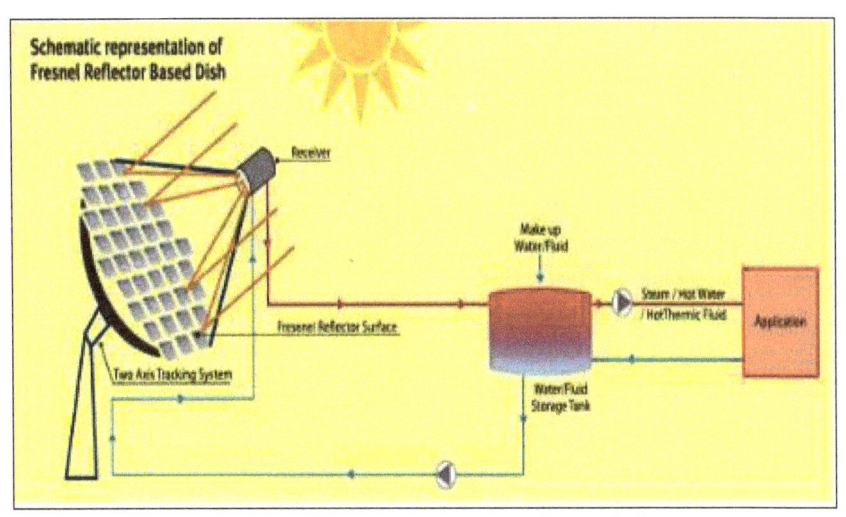

Schematic diagram for fresnel reflector dish.

The reflector and the receiver mounted on top are moved to track the Sun such that the reflectors faces the Sun at all times throughout the year. These systems, because of their size, require careful structural design to withstand their heavy weight and strong winds. Fresnel Dishes have high efficiencies throughout the year.

Schematics of ARUN Dish

Key Design Variant

Presently ARUN Solar Dish is available in three design variant, namely.

ARUN 160   ARUN 100   ARUN 30

Design variant of arun dish.

Paraboloid Dish

Underlying Principle

Paraboloid Dish consists of mirrors mounted on a truss structure such that the incident sunlight is reflected on to a cavity receiver which is specially designed to reduce convective and radiation heat losses. The receiver which is insulated on the outside is held in a fixed position in relation to the reflectors by means of a suitable structure. The entire array of mirrors and receiver move to track the sun. Cavity receivers allow energy to be intercepted by a small aperture or opening which results in low losses. The inner part of the cavity is specially coated to increase its absorption of the sunlight that falls on it, and there are mechanisms to prevent or reduce convective heat losses from the receiver.

The reflector and receiver mounted on top are moved to track the sun such that the reflector faces the sun at all times throughout the year. These systems have light structures. They can be mounted wherever space permits and therefore are suitable for retrofitting in congested layouts. Paraboloid Dishes have high efficiencies throughout the year.

Schematic of Paraboloid Dish

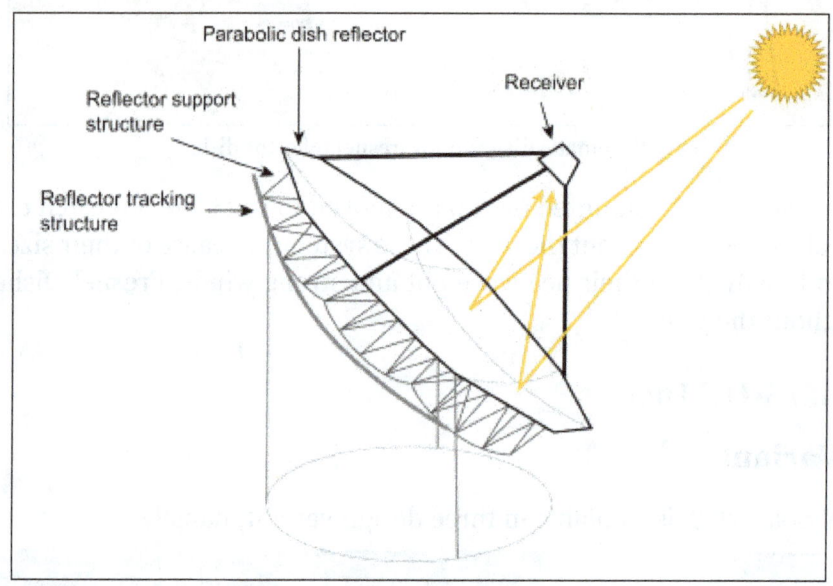

Schematic diagram for paraboloid dish.

Parabolic Trough Concentrator (PTC)

Parabolic Trough Concentrators (PTC) are troughs made from shaped metal and coated with a reflecting material such as highly polished metal (usually aluminium) or metallised plastic which can withstand sunlight as well as rain and the elements. These surfaces reflect the incident sunlight on to a metallic collector pipe (the receiver) that runs axially along the trough. The pipe is specially coated to increase its absorption of the sunlight that falls on it and is encased in a glass tube to reduce convective heat losses from the receiver. Sometimes, to improve the insulation effect, the space between the receiver and the glass envelope is evacuated and the ends are sealed. Several such PTCs can be connected in series on a common axis.

The two common methods of mounting PTCs are with the focal axis:

- In a horizontal E-W direction and the trough is adjusted continuously so that the incident sunlight has the least angle of incidence.

- In the horizontal N-S direction with the troughs being moved to track the sun from east to west from morning to evening. These systems are preferable when a large contiguous area is available and relatively large quantities of heat are needed.

Linear Fresnel Reflector (LFR)

Linear Fresnel Reflecting Concentrator (LFRC) is made from multiple strips of straight reflecting material which are mounted on specially designed frames. The mirrors are arranged in a fresnelized manner which reflects the incident sunlight on to a focal line of metallic collector pipe (the receiver) that runs axially above the array of reflectors. The pipe may be specially coated to increase their absorption of the sunlight that falls on them and is encased in a glass tube to reduce convective heat losses from the receiver. Sometimes, to improve the insulation effect, the space between the receiver and the glass envelope is evacuated and the ends are sealed. The upper portion of the receiver is often insulated to prevent heat loss. The reflector strips are moved such that they always reflect the incident sunlight on to the receiver tube.

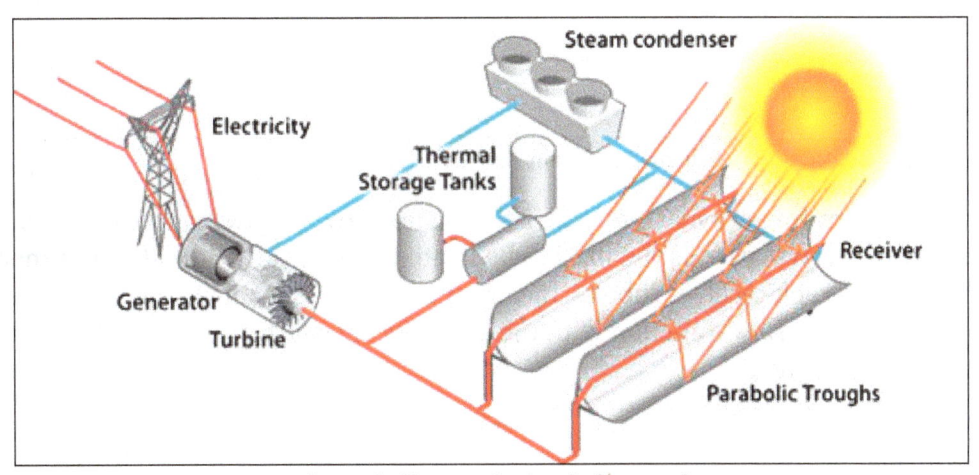

Schematic diagram for parabolic trough.

Several such LFRCs can be connected in series on a common axis. The two common methods of mounting LFRCs are with the focal axis:

- In a horizontal E-W direction and the reflectors are adjusted continuously so that the incident sunlight has the least angle of incidence.

- In the horizontal N-S direction with the troughs being moved to track the sun from east to west from morning to evening. These systems are preferred when a large contiguous area is available and relatively large quantities of heat are needed. They have efficiencies similar to those of PTCs, but can be made more robust.

Schematic

Non-Imaging Concentrator

To make an image, all the parallel light rays should be concentrated at a single point. Non-Imaging collectors are the collectors which can't make the image (of sun).Non-Imaging Concentrators (NIC) are also called Compound Parabolic Collectors (CPC). The NIC consists of a specially coated absorber tube that is enclosed in a concentric glass cover to reduce convection losses. The annulus between the tube and its cover is evacuated. The tube is placed at the focal plane of two reflectors shaped as parabolic troughs. The axes of the two parabolas are inclined at the acceptance angle which is the angle by which direct light may deviate from the normal and yet allow the reflected light to reach the receiver; a high acceptance angle requires fewer tracking adjustments.

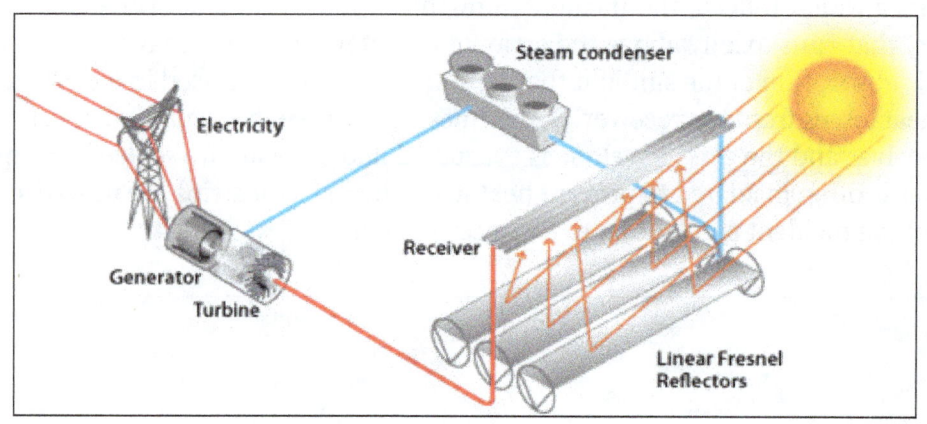

Schematic diagram for linear fresnel system.

Usually such concentrators are made of panels which are 1 to 6 m wide and can go up to 60 m of collector area. The receiver carries the fluid to be heated which could be water or a thermic fluid, and are connected to each receiver tube by means of a header. The most common method of NIC panel mounting is in the E-W direction with the panel facing south and inclined from the ground at an angle of latitude + 10°.

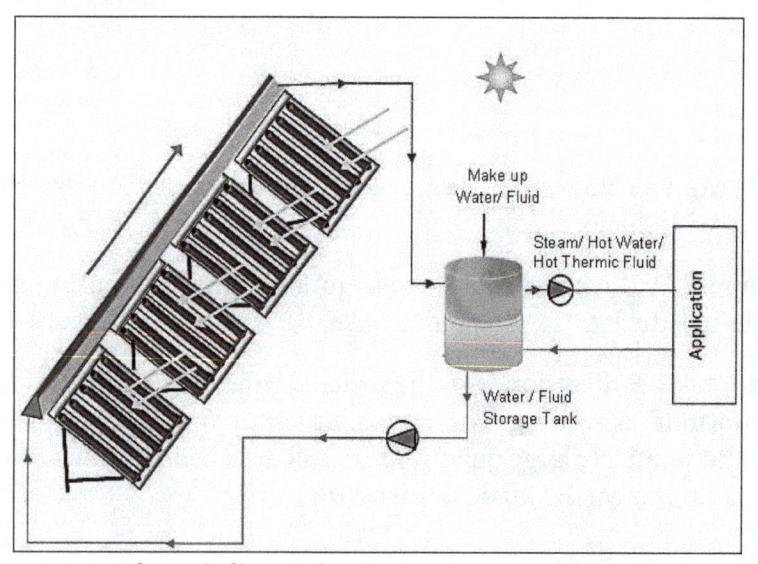

Schematic diagram for non-imaging concentrators.

Concentrated Solar Power (CSP) Technology

Principle of CSP Technology

Concentrated solar power (CSP) is also a solar thermal technology. Here the light energy of the sun is concentrated by using reflective mirrors to generate heat, which in turn produces steam to run a turbine. The generator coupled with the turbine rotates and produces electricity.

Parabolic Trough Collector (PTC)

A parabolic trough collector (PTC) is a linear concentrating system made of long, parabolic-shaped mirrors and a receiver tube placed along the focal axis of the parabola. DNI is concentrated onto the receiver tube, where solar energy is absorbed by the HTF. A glass envelope is often placed around the HCE to limit convection losses and further improve the collector efficiency; the annulus space between the glass envelope and the receiver tube can be under vacuum. Common PTCs achieve concentration ratios of 50, and the HTF temperature can reach up to 400 °C. Parabolic troughs are highly modular and can be arranged in solar fields of various sizes and architecture; however, to minimize losses, the collector axis must be oriented either in an east-west or in a north-south direction, both of which require single-axis tracking. In the case of a smaller solar field, dual-axis tracking can be used to reduce optical losses; however, this is relatively uncommon for linear concentrators.

Parabolic trough collectors are assembled from three main functional units: the concentrator, a mirrored trough with a parabolic-shaped cross-section, uses a tracking device to track the course of the sun so that the incident radiation is concentrated along the focal line on the absorber tube or receiver tube. The collectors are normally aligned in a north-south orientation so that shortly after sunrise the light from the low-lying sun in the east is almost vertically incident on the parabolic opening, the so-called aperture. During the course of the day the sun (in the northern hemisphere) moves southwards and its light falls at an increasingly oblique angle on the collector. The sunrays continue to focus through the tracking parabolic aperture onto the absorber tube but are reflected back across a longer distance. The inclined incident sunlight causes the radiation energy captured per unit area to be correspondingly reduced relative to the cosine of the incident angle (cosine effect). In addition, the inclined reflected sunrays on the north end of the collector miss the absorber tube; this is known as end loss. In order to reduce the relative proportion of these end losses, the parabolic troughs are constructed to be as long as possible. Because the solar radiation during the mornings and evenings is weakened relative to midday as a result of the longer path through the atmosphere, on a cloudless day the usable energy per collector surface is fairly evenly spread across the day.

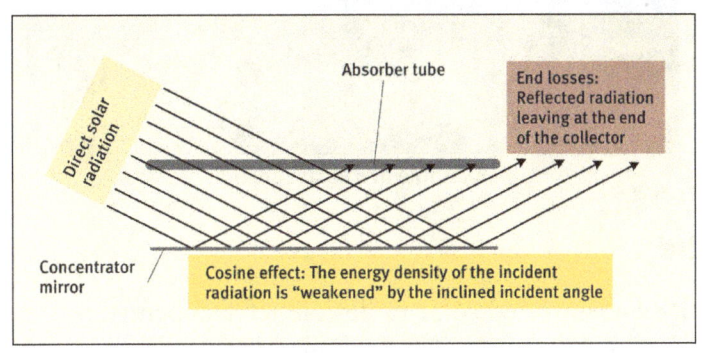

Cosine effect and end losses.

Highly Concentrated

The most important properties for an efficient concentrator are a highly specular reflectance for light of all wavelengths in the solar spectrum as well as a precise parabolic shape. Specular reflectance means that as many rays as possible are reflected according to the "angle of incidence equals the angle of reflection" law and as few rays as possible are absorbed or scattered. Deviations in the parabolic shape cause the radiation to "miss" the absorber tube.

Tracking the Sun

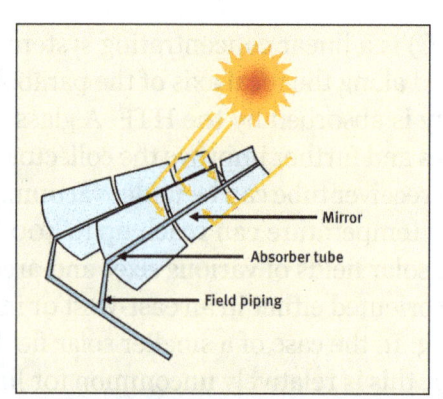

How the parabolic trough works.

Only radiation that is vertically incident to the optical axis is concentrated on the focal point. This is why the concentrator has to continually track the path of the sun. Hydraulic drive systems are predominantly used for this purpose; smaller collectors also use electric motors. The drive systems are controlled either with sensors that determine the position of the collectors relative to the sun's elevation, or by numerically calculating the sun's elevation and the position sensors for the collectors, or by using a combination of both.

Example of a solar sensor.

The Absorber Tube

In the absorber tube the concentrated solar radiation is transformed into heat and transferred to the heat transfer medium flowing inside. The steel tube is coated with an optically selective coating that maintains high absorbance in the solar spectrum wavelength range but high reflectance in the infrared spectrum, i.e. it emits as little as possible. These days, absorption rates of 96 % and emission rates of just 9 % are achieved. In order to prevent heat losses to the ambient air, the absorber tube is surrounded by an evacuated glass envelope tube. The different heat expansion in the glass and metal tube during operation, with temperatures of up to 500 °C, is balanced out using metal bellows at the tube ends. Low-iron glass and anti-reflective coatings ensure that the concentrated radiation can penetrate through the glass tube and strike the absorber coating with as little loss as possible, whereby 96 % transmission rates are achieved.

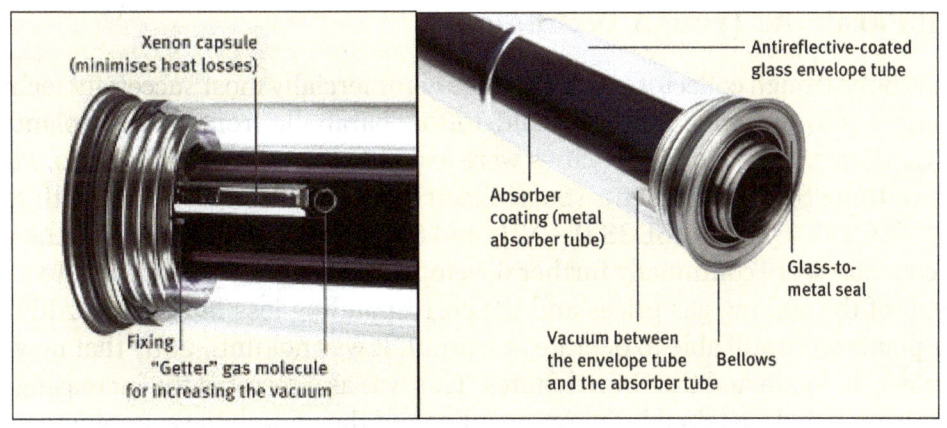

The absorber tube is a key component of parabolic trough collectors.

The Collector System

Example of a Eurotrough with torque boxes.

Parabolic trough collectors have a modular structure. The concentrator collectors are mounted in rows on supporting pylons and are connected with one another so that they are torsionally stiff. Since the capturable soar power is proportional to the concentrator surface area, the collector structure is aimed at using as few components as possible, which include drive systems, movable tube connections on the collector ends and absorber tubes. An extension of the collector length increases the surface area per drive system but requires torsionally stiff structures to transfer the superimposed weight and wind forces from the modules to the drive unit without causing performance-reducing deformations to the concentrator. Increasing the aperture width enables the number of absorber tubes per unit area to be reduced. However, as the surface area increases, so does the wind load, which needs to be taken into account when designing the structures.

Example of a LS3 collector with a space frame.

Successful Parabolic Trough Technology

Until now parabolic trough collectors have been the commercially most successful technology used for solar thermal power plants. Since the mid-1980s, parabolic trough power plants have been operated in California whose total capacities were expanded to 354 MWel by 1990. Parallel to the construction of these SEGS-type plants (Solar Electricity Generating Systems) with rated capacities of 14 MW (SEGS I), 30 MW (SEGS II – VII) and 80 MW (SEGS VIII and IX), the components and system concepts were continually further developed. This development came to a preliminary end as a result of the sinking gas prices and the corresponding loss in revenue. Although the existing power plants were still able to operate at a profit, it was not until 2007 that new plants were built, particularly in Spain and the United States. This was as a result of the increasing importance of climate protection and sustainable energy supplies and the underlying economic conditions that were correspondingly created.

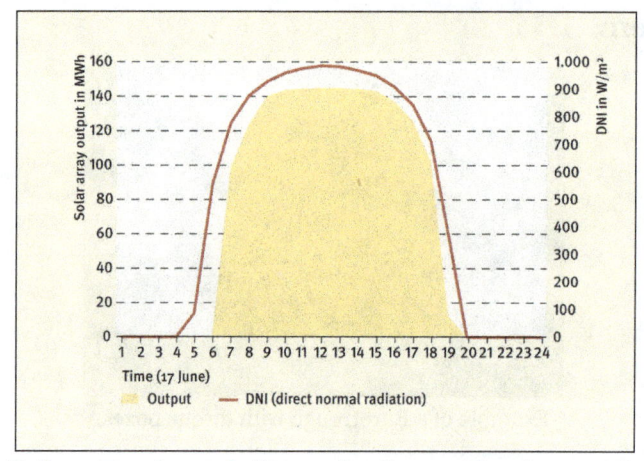

Daily course of a parabolic trough collector using the example of a 50-MW power plant at the Barstow site (California).

Higher Process Temperature – Greater Efficiency

In addition to the aforementioned optimisation of the collector system, further potential is provided by transforming the heat collected in the collector array into electricity as efficiently as possible through increasing the upper process temperature for the conversion process. Since the operating temperature of the predominantly used thermal oil is limited to just under 400 °C for thermal stability reasons, other heat transfer media are being investigated for their suitability. The most progress has been made with developments concerned with the use of direct steam generation in the collector array and molten salt as the heat transfer medium. Both have specific advantages and challenges.

Solar Direct Steam Generation

Direct steam generation enables not only the process temperature to be optimised but also saves on all components in the thermal oil circuit including the associated costs and efficiency losses. The technological challenge here is that the entire tube system needs to be designed for the high pressures of about 100 bar that are desired for turbine operation. The two-phase flow also presents particular challenges in terms of the controllability and thermo-mechanical loads. The heat injection first of all creates small steam bubbles that collect to form larger bubbles above the liquid phase. As a result of the larger specific volume of the steam, the expanding steam bubbles accelerate to form so-called water slugs.

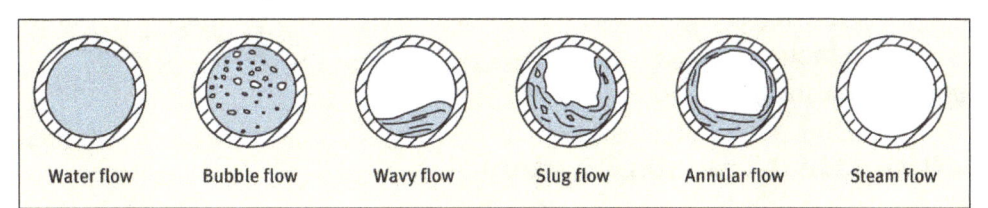

The illustration shows the possible flow forms that can occur in horizontal tubes during evaporation.

Because the heat transfer between the tube wall and the steam phase is considerably worse than between the tube wall and the fluid, the tube temperature increases after this so-called dry-out point. This can cause considerable thermal stress to the tube, since the dry-out point can very rapidly shift with fluctuating solar radiation. The temperature distribution in the tube cross-section can occur in the morning or evening. There is a clear temperature difference between the outer heated and the inner unwetted tube surfaces and the unheated moist section opposite.

In order to utilise the advantages of direct steam generation and simultaneously avoid the risks caused by undesired operating conditions, various process concepts have been developed: a feature of the forced oncethrough process is its simple structure. The water fed into the collector string is preheated, completely evaporated and then superheated in one passage. With the recirculation concept, rather than evaporating the entire injected water, a water-steam mixture is fed into a pressure vessel and separated there by means of gravity. This concept is very robust but incurs higher costs as a result of the necessary pressure vessel, the recirculation pump and losses in the recirculation system.

The first commercial parabolic trough power plant that dispenses with thermal oil and generates the steam directly in the absorber tubes commenced operation in Kanchanaburi (Thailand) at the end of 2011. The plant, which has a rated capacity of 5 MW, works at 30 bar and 330 °C. It is planned to increase both the plant size and the fresh steam parameters in already planned follow-up projects. The aerial view of the plant clearly shows the division of a larger sub-array for pre-heating and steam generation and a smaller superheater array.

Kanchanaburi solar thermal power plant in Thailand with direct steam generation.

Because cheap steam storage systems are not yet available, solar direct steam generation is initially most suitable for smaller systems or hybrid power plants that combine solar and fossil energy. The ability to adapt the steam conditions to the respective feed-in point enables the proportion of solar energy to be increased when integrated with combined cycle power plants. Concepts are also being developed for saving fuel and correspondingly avoiding CO_2 emissions in coal-fired power plants by integrating solar steam generation.

Molten Salt as a Heat Transfer Medium

Molten salt has already proved itself as a storage medium on a commercial scale. The use of similar salts as a heat transfer medium not only leads to considerable savings in the plant technology but also allows the operating temperature to be increased relative to thermal oil. This enables greater process efficiencies and higher storage capacities for the same storage volume.

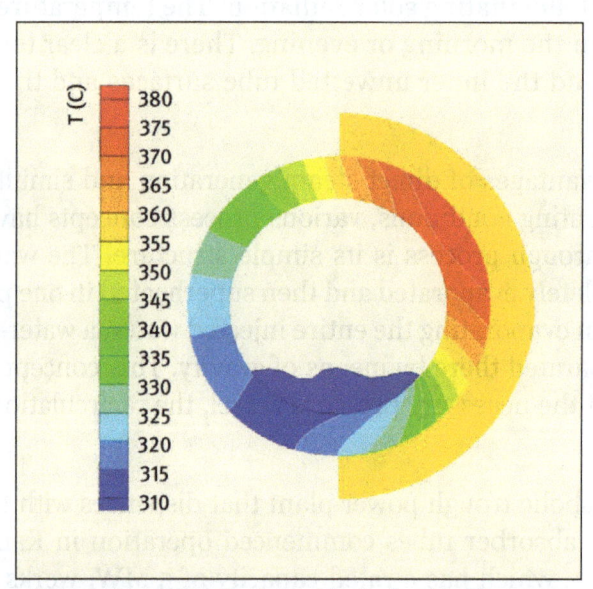

Temperature distribution with single-sided heating.

The most commonly used salt is a mixture of sodium and potassium nitrate. It has a melting point of 238 °C. In periods without irradiation, special precautions are therefore necessary in order to prevent the salt from freezing. The simplest measure is to permanently recirculate warm salt in order to keep the plant at the right temperature. This incurs corresponding heat losses. For this reason research is being conducted on developing salt mixtures with melting points that are as low as possible. Here it also needs to be ensured that these mixtures remain stable at high temperatures, are not corrosive and are cheap. Nitrate salt mixtures with three or more components look highly promising. Chloride mixtures are also being investigated, but these implicate increased corrosion problems.

In July 2010, ENEL and Archimede Solar launched the first application of this technology at the power plant scale in Priolo Gargallo in Sicily. The collector array has a collector surface area of about 30,000 m² and heats the molten salt to 550 °C. The steam generated with the hot molten salt is fed into the steam circuit belonging to ENEL's neighbouring combined cycle power plant and contributes about 5 MW to the electricity generation. System studies show that there are particular economic advantages for plants with molten salt with large rated capacities (> 150 MWel) and storage systems with capacities of 10 to 12 full-load hours.

Storage Concepts

Heat storage systems have two main tasks in solar thermal power plants: on the one hand, relatively small storage systems can be used to balance out fluctuating irradiation and facilitate stable plant operation; on the other hand, large storage systems can extend the operating time of power plant blocks beyond the periods when there is irradiation or enable them to be completely decoupled. The storage size is often specified in full-load hours: that is the number of hours in which the power plant could be operated at full power solely from the storage system. Buffer storage systems typically have a capacity of ½ to 1 full-load hour, whereas large storage systems with 8 to 12 full-load hours enable power plant operation around the clock.

Commonly used are storage systems in which molten salt from a "cold" tank (about 290 °C) is heated to around 390 °C in a heat exchanger using heating oil from the solar array and then stored in a hot tank. The storage capacity depends on the temperature difference between hot and cold, the volume of the tank and the so-called specific thermal capacity of the storage medium. This is the thermal volume that a kilogram of the storage medium can absorb per degree increase in temperature. Since the storage costs are substantially determined by the volume of the storage material, storage media that are as cheap as possible and with a high thermal capacity are used, whereby the temperature difference between the charged and discharged conditions should be as high as possible.

With steam generation, a large proportion of the heat is required at a constant temperature for the phase change from liquid water to steam. That is why a storage system is desirable here which can also absorb and release heat at a constant temperature level. This property is provided by latent heat storage systems that use energy released in the phase change from liquid to solid. Such storage systems are charged using condensating steam. The released heat is transferred to the solid salt and begins to melt this. In the charged condition, the entire salt in the storage system is liquid. To discharge the system, water is injected which extracts the heat from the molten salt so that this gradually hardens while the water evaporates. A considerable challenge when designing such storage systems is the fact that the hardening salt forms a crust on the evaporation tube that conducts heat poorly and so hinders an efficient discharge.

Linear Fresnel Reflector (LFR)

Linear Fresnel reflector (LFR) also based on solar collector rows or loops. However, in this case, the parabolic shape is achieved by almost flat linear facets. The radiation is reflected and concentrated onto fixed linear receivers mounted over the mirrors, combined or not with secondary concentrators. One of the advantages of this technology is its simplicity and the possibility to use low cost components. Direct saturated steam systems with fixed absorber tubes have been operated at an early stage of use of LFR technology. This technology eliminates the need for HTF and heat exchangers. Increasing the efficiency depends on superheating the steam. Superheated steam up to 500 °C has been demonstrated at pilot plant scale and first large commercial superheated LFR plants have begun recently their operation.

Typical large scale LFR system.

LFR plant is highly modular, ranging from a few hundred kW to several MW in size, and offers the lowest land occupancy compared to other CSP technologies. Parabolic troughs represent the optimal solution for achievable concentration ratio and achievable energy yield per aperture area and, hence, the best overall plant efficiency in line-focussing mirror systems. For that reason CSP development efforts have concentrated on parabolic trough geometry.

Nevertheless, in the last decade LFR systems have aroused an increasing interest. The main reason for this is the search for cheaper solar field solutions. The considerable economic advantages of Fresnel collectors are principally related to their constructive simplicity. In addition, Fresnel solar fields permit higher land use efficiency than any other type of solar fields. These advantages can offset the lower solar-to-heat efficiency, and LFR power plants represent an interesting alternative to parabolic trough power plants.

The main advantage of LFR systems is that their simple design of flat or flexibly bent mirrors and fixed receivers requires lower investment costs and offers a wide range of configurations. Originally designed for low and medium power applications, LFRs are now being designed for higher temperatures which facilitate direct steam generation (DSG) which can be used efficiently in the industrial or power generation sector.

Fresnel systems can be configured to operate over a wide range of temperatures, from 200 to 500 °C, although systems with temperatures as high as 550 °C are under development. Applications range from industrial process heat, distributed power generation using the organic Rankin cycle to steam turbine systems.

Developments in energy efficiency in both industry and power generation have focused attention on medium and high temperature solar thermal systems. LFRs have great potential in southern Africa due to the low cost and high percentage of local manufacture inherent in the technology.

Principle of Fresnel mirror system

Principle of Fresnel mirror system.

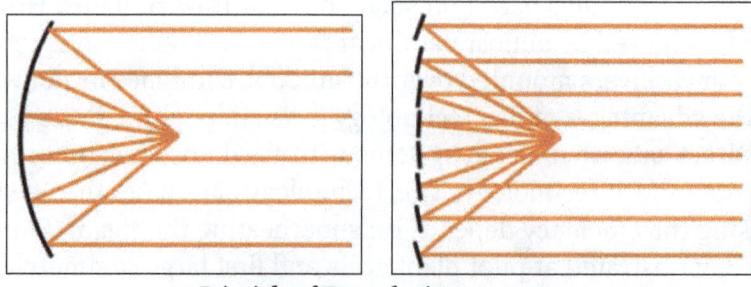

The linear Fresnel reflector technology receives its name from the Fresnel lens, which was developed by the French physicist Augustin-Jean Fresnel for lighthouses in the 18th century. The principle of this lens is the breaking of the continuous surface of a standard lens into a set of surfaces with discontinuities between them. This allows a substantial reduction in thickness (and thus weight and volume) of the lens, at the expense of reducing the imaging quality of the lens. Where the purpose is to focus a source of light this impact on the image quality is not of major importance.

The principle of dividing an optical element into segments which have the same (or a very similar) optical effect as the original optical element can also be applied to mirrors. It is possible to divide a parabolic mirror into annular segments, forming a circular Fresnel mirror which focuses the light that arrives in rays parallel to the optical axis onto the focal point of the paraboloid mirror.

In an analogous way, a linear Fresnel mirror can be constructed substituting a parabolic trough by linear segments that focus the radiation that arrives in a plane parallel to the symmetry plane of the parabolic trough onto the focal line of the parabolic trough. A LFR has a similar effect to a parabolic trough, when considering the concentration of the radiation in a focal line i.e. a LFR behaves like a parabolic trough with the same focal length and the same aperture.

The mirrors focus the sun onto a receiver which contains the heat transfer medium which could be water, oil or even molten salt in some designs. The heat transfer medium used will depend on the operating temperature of the system. The main difference between the two systems lies in the way that the sun's rays are tracked, and this is what gives rise to the cheaper cost of Fresnel.

In the trough system the whole structure rotates about an axis coincident with the focal point of the trough. This means that the mirrors and the collector are connected mechanically, requiring bearings through which the collector tube must pass. In the Fresnel system the individual mirrors rotate to track the sun. There is no mechanical connection between the mirrors and the collector.

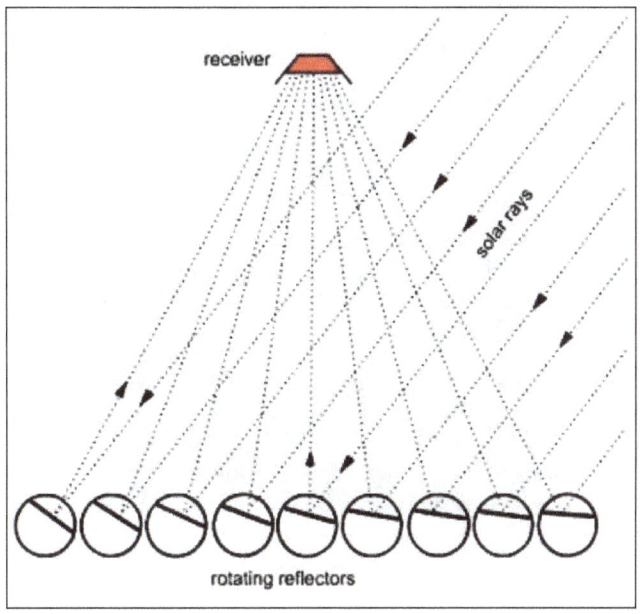

Mirror tracking with Fresnel system.

Separation of the mirrors and the receiver allows high temperature heat transfer mediums to be used, and also a much wider scope in the design of the receiver. The Fresnel system also allows individual control of each mirror, effectively changing the configuration of the reflector to optimise its function.

LFR Construction

Aperture

The mirror aperture is the area of the reflector mirrors in the horizontal position, and defines the amount of solar radiation collected by the LFR. Aperture is usually quoted on a module or single mirror length basis. Increasing the mirror aperture increases the amount of solar radiation reflected to the receiver module. Typical aperture widths for large systems are of the order of 15 m.

The receiver aperture is the area of the receiver per mirror or module length, and varies according to the design of the system. Receiver width is close to the mirror width, but may be less if curved mirrors are used, or wider if other focusing methods are used.

Mirror Construction

Mirrors may be flat or elastically curved and are generally constructed of glass with a composite/metal backing, although other materials are finding their way into the sector. Adding curvature to the mirror increases the concentration ratio and makes the design of the receiver simpler. Mirror width and length will depend on the design. The mirrors are mounted in the tracking system in several different ways. A typical mounting would be a circular loop driven by a tracking motor.

Mounting and tracking system.

Receiver Construction

The receiver generally consists of a secondary reflector mounted above the receiver tube which contains the heat transfer fluid. The receiver may consist of a single tube or several tubes, which may be contained in a vacuum glass tube enclosure.

The LFR system alone cannot reach the same radiation concentration as a parabolic trough. The sun changes its position in relation to the optical axis plane of the system. It is theoretically impossible to design the curvature of the individual mirror strips in such a way that there is always a sharp focal line for parallel radiation, and it is necessary to mitigate the unavoidable optical inaccuracy of the Fresnel collector. This can be done by a secondary concentrator that is located above the receiver tubes.

Concentration Ratio

The collector concentration ratio is the ratio of mirror aperture to receiver aperture. The low-profile setting of LFR collectors maximises the concentration ratio, which enables high temperature output.

Fundamentally, increasing the mirror aperture allows more sunlight to be reflected onto the receiver. Because the low-profile architecture provides for great flexibility in the selection of length and concentration ratio, linear Fresnel collectors can be readily tailored for different target temperatures to meet varying application needs, thus providing practical versatility of usage. As an example, the Fresdemo system has a primary mirror field width of 21 m and a receiver width of 0,5 m, giving a concentration ratio of 42.

Heat Transfer Fluids

The fixed design of the receiver gives a wide choice of heat transfer fluids. Water is used as the transfer fluid at low temperatures, oil as temperatures increase, steam for much higher temperatures and molten salt for the highest ranges. In addition, pressurised CO_2 or air are being considered for higher temperature operation.

Conversion Efficiency

LFR exhibit optical conversion efficiency in the region of 65%. Conversion efficiency depends on the angle of incidence of the sun, and thus average efficiency will vary with latitude. Thermal peak output of 562 W/m² in terms of primary reflector aperture area, and 375 W/m² in terms of installation area usage, is claimed for a typical commercially available system.

Secondary concentrator on receiver.

Applications

Medium Heat Generation

Medium heat systems for industrial applications or supplementary power systems operate in the temperature range 100 to 250 °C and may use water or oil as the heat transfer medium. There are numerous systems of this type in operation.

Direct Steam Generation

Thanks to the fixed absorber tubes, direct steam generation (DSG) is easier in LFR power plants than in parabolic trough power plants. DSG has several advantages over oil heat transfer:

- Steam as heat transfer fluid allows higher temperatures because there is no danger of thermo oil cracking. Novatec Solar is planning a new Fresnel power plant generation that operates at 450 °C.

- The number of construction components can be reduced because no heat exchange has to be realised between a solar field heat transfer fluid/thermo oil) and Rankine cycle working fluid (water/steam).

- The thermo oil itself is an expensive component of CSP plants, so the lack of thermo oil is a direct advantage.

- As there is no heat transfer between two heat transfer fluids, there is one thermal loss factor less.

- The use of steam as a heat transfer fluid may reduce the mean heat transfer fluid temperature in the absorber tube (even at higher final temperatures) and reduce thermal losses. This reduction is possible because in a large part of the receivers the boiling process is realised, which takes place at a reduced temperature. Only in the small part, where the superheating of the steam is realised (if there is superheating), high temperatures are reached.

- Water has further advantages in comparison to other HTFs: It is environmentally friendlier than thermo oil so that leakages in a steam generating plant do not produce environmental dangers.

Secondary reflector multiple tube technology (Ausral).

Water is less corrosive than salt. Its freezing temperature is much lower than the freezing temperature of salt and even slightly lower than of thermo oil. The effort required to ensure adequate anti-freeze protection is reduced significantly.

There are several systems on the market that use DSG. In this system water is fed through pipes in the receiver where it is converted to steam and then to superheated steam. Steam is used to drive turbines directly or as supplementary heating for thermal power plants. Steam is also used for industrial purposes.

Molten Salt Systems

High temperature systems (≈550 °C) using molten salt as the heat transfer medium have been developed. Such systems incorporate heat storage systems for prolonged operation. Figure shows a high temperature plant using salt as the heat transfer medium.

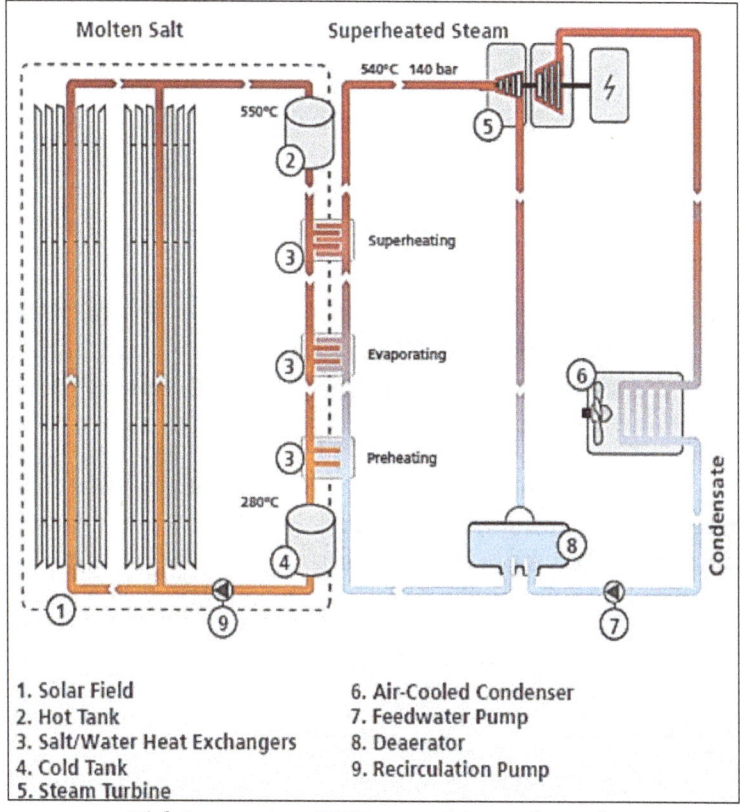

High temperature CLF system using molten salt.

Because of the simplicity of design and the high operating temperatures now possible, LFR are being considered as alternatives to tower and heliostat based CSP systems. In addition, the short optical path between mirror and receiver eliminates many of the tracking problems as well as environmental objections associated with tower/heliostat based systems. In addition LFR salt based systems are able to recover from "salt freezing".

Supplementary Heat or Steam Generation for Power Plants

Fresnel mirrors are being considered as a means to supplement heat in thermal power stations.

Parts of the cycle such as preheating of the boiler feed water use low to medium temperature heat sources, and solar thermal systems can provide this heat. Typical applications are to substitute the bleed steam used for preheating boiler feedwater as shown in figure.

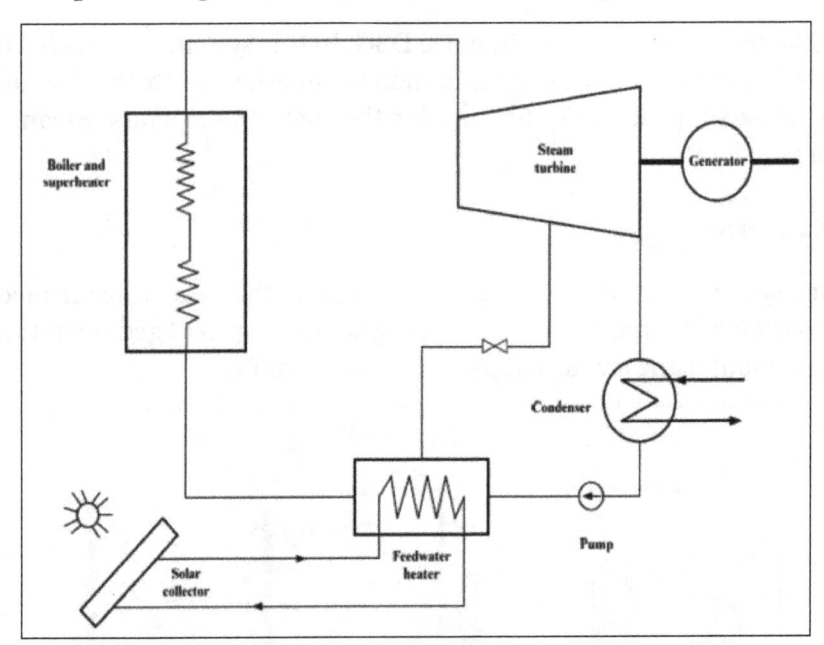

Single-stage regenerative Rankine cycle with open feedwater heater.

The solar heat source is connected in parallel with the bled steam source and when available provides the heat normally used from the steam. The increased steam output then enables additional power generation from the turbine (solar boosting mode) or fuel consumption can be reduced (fuel saver mode).

More complex systems use several stages of regeneration with heat substitution at each stage. Heat may be stored for periods when the solar resource is low, or the plant may be operated in a two shift mode. Using solar power may have two effects:

- Increase in the power generation capacity of the plant during periods when solar power is available.

- Decrease in coal consumption with no increase in power generation.

The integration of solar thermal collectors into conventional fossil plants, or solar aided power generation (SAPG), has proven a viable solution to address the intermittency of power generation and combines the environmental benefits of solar power plants with the efficiency and reliability of fossil power plants. In SAPG technology, thermal oil can be used as solar heat carrier and no solar steam needs to be generated, therefore the pressure of solar system can be much lower than that of the solar collector using water/steam as the heat carrier. Newer developments make use of the higher heat content of solar steam directly, provided from DSG plants.

The temperature of the heat source is one of the major defining factors of a power plant – higher temperature results in higher overall power plant efficiency. With SAPG, the heat source temperature is not limited by the solar input temperature and is therefore an effective means of utilising low or medium solar heat (250 °C) for power generation. However, internationally the adoption of the technology has been slow, despite it being a viable and quick means of CO_2 emission reduction.

A study was conducted on a simulated SAPG power plant at Lephalale, which was based on a generic 600 MW subcritical fossil power plant with a reheater and regenerative Rankine cycle with two low pressure feedwater heaters (FWH) downstream of the deaerator and three high-pressure FWHs upstream.

This study showed that integrating solar thermal with steam plant is approximately 1,5 times as efficient at converting solar energy to electricity as a CSP stand-alone generating plant. Therefore, a solar assisted high pressure feedwater heater system at an existing coal-fired power station is 1,8 times more cost-effective than a stand-alone CSP plant. The conversion efficiency is claimed to be of the same order as solar PV.

Compact Linear Fresnel Reflector

A compact linear Fresnel reflector (CLFR) – also referred to as a concentrating linear Fresnel reflector – is a specific type of linear Fresnel reflector (LFR) technology. They are named for their similarity to a Fresnel lens, in which many small, thin lens fragments are combined to simulate a much thicker simple lens. These mirrors are capable of concentrating the sun's energy to approximately 30 times its normal intensity.

Linear Fresnel reflectors use long, thin segments of mirrors to focus sunlight onto a fixed absorber located at a common focal point of the reflectors. This concentrated energy is transferred through the absorber into some thermal fluid (this is typically oil capable of maintaining liquid state at very high temperatures). The fluid then goes through a heat exchanger to power a steam generator. As opposed to traditional LFR's, the CLFR utilizes multiple absorbers within the vicinity of the mirrors.

The first linear Fresnel reflector solar power system was developed in Italy in 1961 by Giovanni Francia of the University of Genoa. Francia demonstrated that such a system could create elevated temperatures capable of making a fluid do work. The technology was further investigated by companies such as the FMC Corporation during the 1973 oil crisis, but remained relatively untouched until the early 1990s. In 1993, the first CLFR was developed at the University of Sydney in 1993 and patented in 1995. In 1999, the CLFR design was enhanced by the introduction of the advanced absorber. In 2003 the concept was extended to 3D geometry. Research published in 2010 showed that higher concentrations and / or higher acceptance angles could be obtained by using nonimaging optics to explore different degrees of freedom in the system such as varying the size and curvature of the heliostats, placing them at a varying height (on a wave-shape curve) and combining the resulting primary with nonimaging secondaries.

Design

Reflectors

The reflectors are located at the base of the system and converge the sun's rays into the absorber. A key component that makes all LFR's more advantageous than traditional parabolic trough mirror systems is the use of "Fresnel reflectors". These reflectors make use of the Fresnel lens effect, which allows for a concentrating mirror with a large aperture and short focal length while simultaneously reducing the volume of material required for the reflector. This greatly reduces the system's cost since sagged-glass parabolic reflectors are typically very expensive. However, in recent years thin-film nanotechnology has significantly reduced the cost of parabolic mirrors.

A major challenge that must be addressed in any solar concentrating technology is the changing angle of the incident rays (the rays of sunlight striking the mirrors) as the sun progresses throughout the day. The reflectors of a CLFR are typically aligned in a north-south orientation and turn about a single axis using a computer controlled solar tracker system. This allows the system to maintain the proper angle of incidence between the sun's rays and the mirrors, thereby optimizing energy transfer.

Absorbers

The absorber is located at the focal line of the mirrors. It runs parallel to and above the reflector segments to transport radiation into some working thermal fluid. The basic design of the absorber for the CLFR system is an inverted air cavity with a glass cover enclosing insulated steam tubes, shown in figure. This design has been demonstrated to be simple and cost effective with good optical and thermal performance.

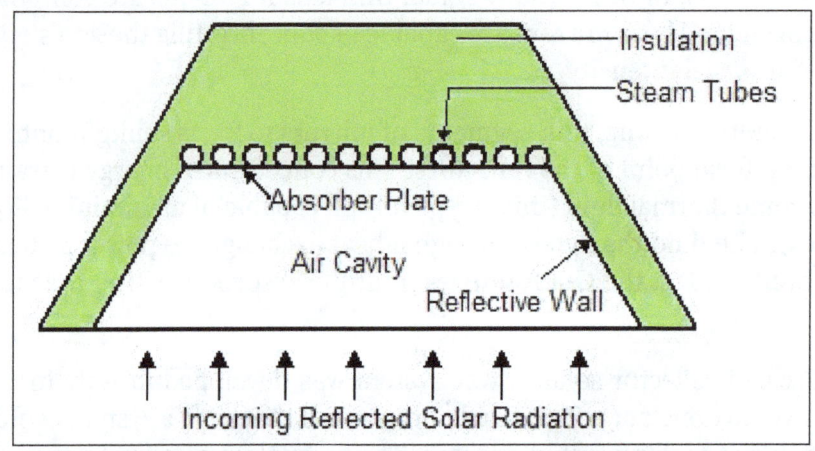

Incident solar rays are concentrated on insulated steam tubes to heat working thermal fluid.

For optimum performance of the CLFR, several design factors of the absorber must be optimized.

- First, heat transfer between the absorber and the thermal fluid must be maximized. This relies on the surface of the steam tubes being selective. A selective surface optimizes the ratio of energy absorbed to energy emitted. Acceptable surfaces generally absorb 96% of incident radiation while emitting only 7% through infra-red radiation. Electro-chemically deposited black chrome is generally used for its ample performance and ability to withstand high temperatures.

- Second, the absorber must be designed so that the temperature distribution across the selective surface is uniform. Non-uniform temperature distribution leads to accelerated degradation of the surface. Typically, a uniform temperature of 300 °C (573 K; 572 °F) is desired. Uniform distributions are obtained by changing absorber parameters such as the thickness of insulation above the plate, the size of the aperture of the absorber and the shape and depth of the air cavity.

As opposed to the traditional LFR, the CLFR makes use of multiple absorbers within the vicinity of its mirrors. These additional absorbers allow the mirrors to alternate their inclination, as illustrated in figure. This arrangement is advantageous for several reasons.

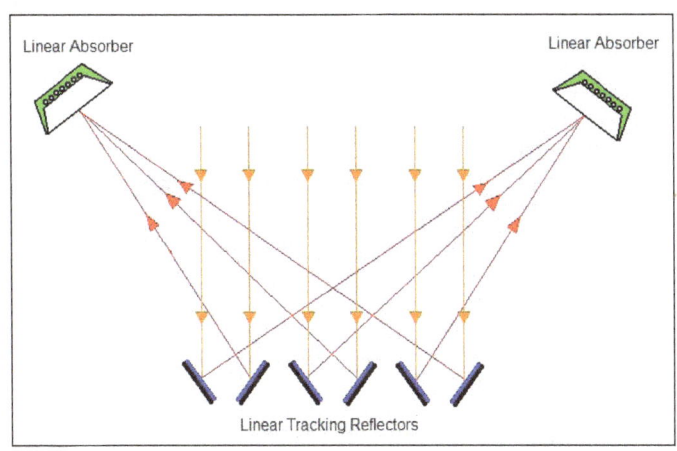

CLFR solar systems alternate the inclination of their mirrors to focus solar energy on multiple absorbers, improving system efficiency and reducing overall cost.

- First, alternating inclinations minimize the effect of reflectors blocking adjacent reflectors' access to sunlight, thereby improving the system's efficiency.

- Second, multiple absorbers minimize the amount of ground space required for installation. This in turn reduces cost to procure and prepare the land.

- Finally, having the panels in close proximity reduces the length of absorber lines, which reduces both thermal losses through the absorber lines and overall cost for the system.

Applications

Areva Solar (Ausra) built a linear Fresnel reflector plant in New South Wales, Australia. Initially a 1 MW test in 2005, it was expanded to 5MW in 2006. This reflector plant supplemented the 2,000 MW coal-fired Liddell Power Station. The power generated by the solar thermal steam system is used to provide electricity for the plant's operation, offsetting the plant's internal power usage. AREVA Solar built the 5 MW Kimberlina Solar Thermal Energy Plant in Bakersfield, California in 2009. This is the first commercial linear Fresnel reflector plant in the United States. The solar collectors were produced at the Ausra factory in Las Vegas. In April 2008, AREVA opened a large factory in Las Vegas, Nevada to produce linear Fresnel reflectors. The factory was planned to be capable of producing enough solar collectors to provide 200 MW of power per month.

In March 2009, the German company Novatec Biosol constructed a Fresnel solar power plant known as PE 1. The solar thermal power plant uses a standard linear Fresnel optical design (not CLFR) and has an electrical capacity of 1.4 MW. PE 1 comprises a solar boiler with mirror surface of approximately 18,000 m2 (1.8 ha; 4.4 acres). The steam is generated by concentrating sunlight directly onto a linear receiver, which is 7.40 metres (24.28 ft) above the ground. An absorber tube is positioned in the focal line of the mirror field where water is heated into 270 °C (543 K; 518 °F) saturated steam. This steam in turn powers a generator. The commercial success of the PE 1 led Novatec Solar to design a 30 MW solar power plant known as PE 2. PE 2 has been in commercial operation since 2012.

From 2013 on Novatec Solar developed a molten salt system in cooperation with BASF. It uses molten salts as heat transfer fluid in the collector which is directly transferred to a thermal energy storage. A salt temperature of up to 550 °C (823 K; 1,022 °F) facilitate to run a conventional steam turbine for Electricity generation, Enhanced oil recovery or Desalination. A molten salt demonstration plant was realized on PE 1 to proof the technology. Since 2015 FRENELL GmbH, a management buy-out of Novatec Solar took over the commercial development of the direct molten salt technology.

Solar Fire, an appropriate technology NGO in India, has developed an open source design for a small, manually operated, 12 kW peak Fresnel concentrator that generates temperatures up to 750 °C (1,020 K; 1,380 °F) and can be used for various thermal applications including steam powered electricity generation.

Heliostat Field Collectors (HFCs)

Heliostat field collectors (HFCs) also known as "Power Tower" or "Central Receiver Plant". Heliostats are named helio for sun and stat for the fact that the reflected solar image is maintained at a stationary position throughout the day. They are nearly flat mirrors (some curvature is required to focus the sun's image) that collect and concentrate the solar energy on a tower-mounted receiver located 100 to 1000 meters distant. To maintain the sun's image on the solar receiver, heliostats tracks the sun throughout the day.

The concentrated heat energy absorbed by the receiver is transferred to a circulating fluid that can be stored and later used to produce power. Central receivers have several advantages:

- They solar energy is collected optically and transferred to a single receiver point, thus minimizing thermal-energy transport requirements;
- They typically achieve concentration ratios of 100–1500 and so are highly efficient both in collecting energy and in converting it to electricity;
- They can conveniently store thermal energy;
- They are quite large (generally more than 10 MW) and thus benefit from economies of scale.

Each heliostat at a central-receiver facility has from 50 to 150 m² of reflective surface. The heliostats collect and concentrate sunlight onto the receiver, which absorbs the concentrated sunlight, transferring its energy to a heat transfer fluid. The heat-transport system, which consists primarily of pipes, pumps, and valves, directs the transfer fluid in a closed loop between the receiver, storage, and power-conversion systems. A thermal-storage system typically stores the collected energy as sensible heat for later delivery to the power-conversion system. The storage system also decouples the collection of solar energy from its conversion to electricity. The power-conversion system consists of a steam generator, turbine generator, and support equipment, which convert the thermal energy into electricity and supply it to the utility grid.

Schematic of Heliostat Field Collector

Dish Sterling Engine

Dish/engine systems use a parabolic dish of mirrors to direct and concentrate sunlight onto a central engine that produces electricity. The engine is placed at the focus of the parabolic dish. The dish/engine system is a concentrating solar power (CSP) technology that produces smaller amounts of electricity than other CSP technologies—typically in the range of 3 to 25 KW. The two major parts of the system are the solar concentrator and the power conversion unit.

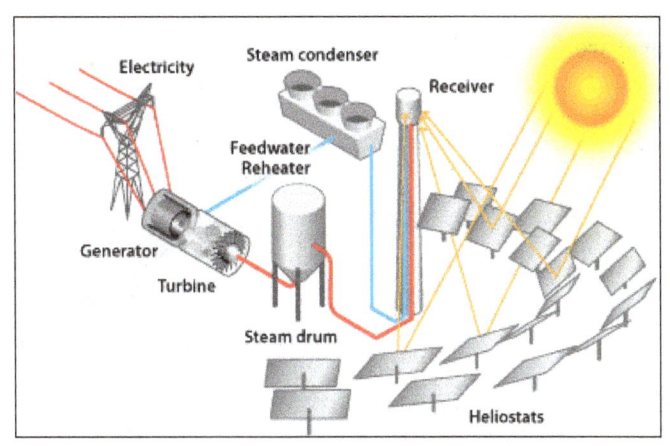

Schematic diagram for heliostat field collector.

Solar Concentrator

The solar concentrator, or dish, collects the solar energy coming directly from the sun. The resulting beam of concentrated sunlight is reflected/focused onto a thermal receiver that collects the solar heat. The dish is mounted on a structure that tracks the sun continuously throughout the day to reflect the highest percentage of sunlight possible onto the thermal receiver.

Power Conversion Unit (Engine)

The power conversion unit includes the thermal receiver and the engine/generator. The thermal receiver is the interface between the dish and the engine/generator. It absorbs the concentrated beams of solar energy, converts the energy to heat, and transfers the heat to the engine/generator. A thermal receiver can be a bank of tubes with a cooling fluid—usually hydrogen or helium—that typically is the heat-transfer medium and also the working fluid for an engine. Alternate thermal receivers are heat pipes, where the boiling and condensing of an intermediate fluid transfers the heat to the engine.

The engine/generator system is the subsystem that takes the heat from the thermal receiver and uses it to produce thermal to electric energy conversion. The most common type of heat engine used in dish/engine systems is the Stirling engine. A Stirling engine uses the heated fluid to move pistons and create mechanical power. The mechanical work, in the form of the rotation of the engine's crankshaft, drives a generator and produces electrical power.

Schematic of Dish Sterling Engine

Efficiency of CST System

Performance of CST system depends on the following factors:

- Direct Normal Solar Irradiance, I_{bn},

- Latitude/Incident effect,
- Required Temperature,
- Wind Velocity,
- Ambient Temperature.

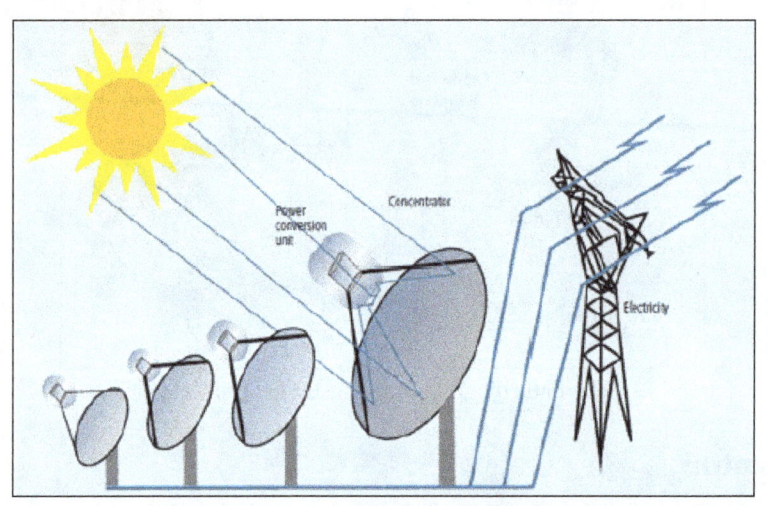

Schematic diagram for dish sterling engine.

Optical Efficiency

Optical efficiency describes the system's ability to absorb radiation that strikes normal to the concentrator aperture. When direct solar radiation reaches the surface of solar collectors, a significant amount of it is lost due to several different factors. The total loss can be divided into three types, which are as follows:

- Optical losses: Solar radiation incident upon the collector that is not converted to heat energy due to following factors:
  - Reflectivity.
  - Transmissivity of the glass tube, τ.
  - Intercept factor, γ, (losses due to tracking, geometry, heat removal factor of fluid).
  - Absorptivity of the absorber selective coating, α.
- Thermal heat losses: Solar radiation that is converted to heat, but lost before it can be used. Losses are due to three modes of heat transfer within the collector:
  - Radiative heat transfer loss.
  - Convective heat transfer loss.
  - Conductive heat transfer loss.
- Geometrical Losses: Due to the incidence angle, θ, of direct solar radiation on the aperture

plane of the collector. The incidence angle of direct solar radiation is a very important factor, because the fraction of direct radiation that is useful to the collector is directly proportional to the cosine of this angle, which also reduces the useful aperture area of the solar collector. The effect of the incidence angle on the optical efficiency and useful aperture area of a PTC is quantified by the incident angle modifier K_{θ_T} and K_{θ_L}, because this parameter includes all optical and geometric losses due to an incidence angle greater than 0°.

Basic Performance Equation

The performance of CST based solar collector operating under steady state conditions can be described by the following equation:

$$\frac{Q}{A_p} = I_{bn} \times \gamma \times \rho(\alpha\tau) \times \cos\theta - a_1 \times (T_m - T_a) - a_2 \times (T_m - T_a)^2$$

The product (ατ) represents cover absorbance of the receiver when receiver is glass enclosed. In case there is no glass enclosure considers τ as 1. The τ represents transmittance of the glass cover. The α represents absorbance of the receiver surface and represents reflectivity of the reflector surface. The γ represents intercept factor which includes the losses due to tracking, geometry, heat removal factor of fluid etc. While the value of ρ, α, τ are properties of material of reflector and receiver and could be known but the value of intercept factor (γ) is not known and could be determined through thermal calculation of the system. The product [γ x ρ(ατ)] is known as optical efficiency η_o.

The terms ρ, α, τ are for normal incidence of solar radiation. But in order to predict the system performance at other incident angles, as would be the case with actual field installations, multiplying factor called the incident angle modifier (K_{θ_T} and K_{θ_L}). Also the effect of angle of incidence on the intensity of beam radiation is introduced in the equation in form of Cos θ where, θ is the angle of incidence.

$$\frac{Q}{A_p} = I_{bn} \times \eta_o \times \cos\theta \times K_{\theta_T} \times K_{\theta_L} - a_1 \times (T_m - T_a) - a_2 \times (T_m - T_a)^2$$

Where, K_{θ_L} is the longitudinal incidence angle modifier and K_{θ_T} is transversal incidence angle modifier (Correction factors). Longitudinal incidence angle modifier K_{θ_L} can be calculated with the angle of incidence caused by declination of sun and transversal incidence angle modifier K_{θ_T} can be calculated with angle of incident due of sun movement from morning to evening.

The thermal efficiency solar collector system is defined as a ratio of the actual useful-energy collected to that of the solar energy intercepted by the solar collector, then the efficiency of the CST based concentrator system may be defined as follows:

$$\eta = \eta_o - a_1 \left(\frac{T_m - T_a}{I_{bn}} \right) - a_2 \times I_{bn} \left(\frac{T_m - T_a}{I_{bn}} \right)^2$$

Thermal Efficiency

The instantaneous efficiency η shall be calculated from the following expression:

$$\eta = \frac{\dot{m} \times C_p (T_{out} - T_{in})}{I_{bn} \times A_p}$$

An appropriate value of C_p at mean fluid temperature shall be used in the above expression. If $\dot{m}$ is obtained from volumetric flow rate measurement, then the density shall be determined for the temperature of the fluid in the flow meter. The instantaneous efficiency shall be presented graphically as a function of $\left(\frac{T_m - T_a}{I_{bn}}\right)$. All the data points shall be plotted along with a statistical curve fitting using the least square.

Computation of Optical Efficiency

The optical efficiency shall be calculated from the following equation:

$$\eta = \eta_o - a_1 \left(\frac{T_m - T_a}{I_{bn}}\right) - a_2 \times \frac{(T_m - T_a)^2}{I_{bn}}$$

For calculating optical efficiency the system is maintained in zero heat loss condition by selecting inlet temperature such that the mean fluid temperature in the receiver lies within ±3 °C of the ambient temperature ($T_m = T_a$). The optical efficiency of the system under zero heat loss condition will be equal to its thermal efficiency.

Table: Efficiency of different technologies.

S. No.	Technology	Tracking	Focus Type	Operating Temperature Range (°C)	Efficiency (%)
1.	NIC	Non Tracking	Line focus	80-140	45-55
2.	Scheffler Dish	Single Axis Tracking	Fixed focus	120-250	30-45
3.	PTC	Single Axis Tracking	Line focus	120-300	40-55
4.	Paraboloid Dish and Arun Dish	Dual Axis Tracking	Point focus	150-400	50-65
5.	LFR	Single Axis Tracking	Line focus	150-300	40-55

Evacuated Tube Collector (ETC)

In flat plate collectors, significant heat is lost mostly by convection and re-radiation through the top surface of the collector. This heat loss increases as the water temperature in the collector gets hotter during the day. So while the collector is highly efficient at the beginning of the day (e.g. 70%), the efficiency decreases as the water circulating through the collector gets hotter.

In evacuated tube systems, this heat loss is reduced by almost totally eliminating conduction and convection heat losses. This is because the space between the absorber and the glass outer tube is evacuated. With little air to move and transfer heat by conduction and convection, heat loss is further reduced. Radiation losses are reduced by incorporating a selective surface on the absorber, similar to flat plate collectors. As a result, evacuated tube collectors can operate at temperatures above 100 °C, compared with about 100 °C for flat plate collectors.

The principle of operation is similar to a flat plate collector in that solar radiation (both direct and diffuse) enters through the glass tube and is absorbed by the absorber plate, which transfers the heat into a heat transfer fluid inside the collector tube.

Solar Pond

A solar pond is a shallow body of water which acts as a solar collector with integral heat storage for supplying thermal energy. There are two types of solar ponds – convective solar pond and non-convective solar pond. The shallow solar pond is a convective solar pond. It consists of a large bag that prevents evaporation but permits convection. The bag has blackened bottom with foam insulation below, and two types of glazing (sheets of plastic or glass) on top. Solar energy heats the water in the bag during the day and at night the hot water is pumped into a large heat storage tank to minimize heat loss. Another type is the deep, saltless pond. Double glazing covers deep saltless pond. When solar energy is not available or at night placing insulation on the top of the glazing reduces heat loss.

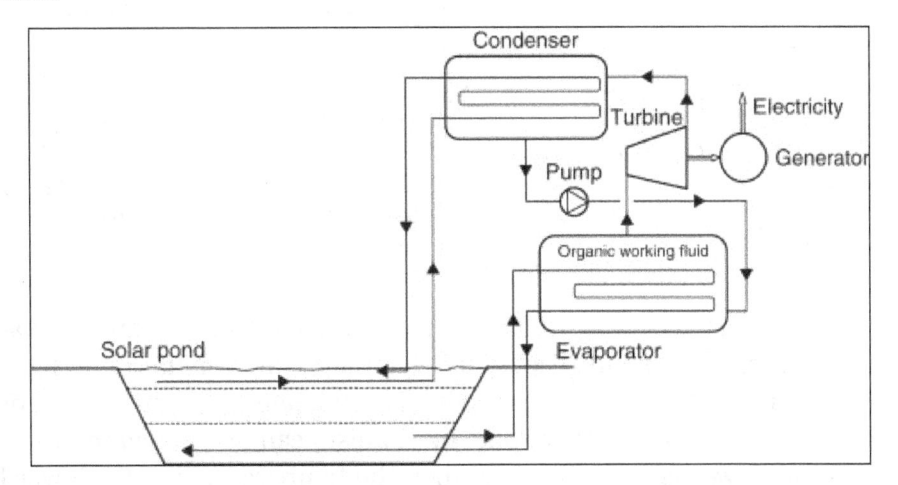

A non-convective solar pond is a large shallow body of water 1 to 5 m deep, but 3-4 m on the average, which is arranged in a way so that the temperature gradient is reversed from the normal. This allows collection of radiant energy into heat (up to 95 °C), storage of heat and transport of thermal energy, at temperature 40-50 °C above normal, out of the system.

There are three types of non-convective solar ponds in terms of the methods of maintaining layered structure. One is SGSP where density gradient is maintained by salt water. The other is membrane solar pond which uses horizontal and vertical membranes. The third one is polymer gel layers solar pond.

A SGSP is a system for solar energy collection and storage. It uses solar radiation to heat water; stores sensible heat in dense saline water; establishes density gradient to prevent convective heat flow and thus stores thermal energy.

A SGSP has 3 main layers. These are UCZ (Upper Convective Zone): top layer; NCZ (Non-convective Zone): middle layer and LCZ (Lower Convective Zone): bottom layer.

UCZ is of almost low salinity and at close to ambient temperature. This zone is the result of evaporation, wind mixing, surface flushing and nocturnal cooling. Generally this layer is maintained as thin (0.3 – 0.5 m) as possible by the use of wave-suppressing meshes or by placing wind-breaks near the ponds.

NCZ is a gradient which is much thicker and occupies 1.5 m or more than half of the depth of the pond. In NCZ, both salt concentration and temperature increases with depth. The vertical salinity gradient in NCZ holds back convection and thus offers the thermal insulation effect. Temperature gradient is formed due to absorption of solar radiation at the pond base. LCZ is a zone of almost constant relatively high salinity (20-30% by weight) at nearly constant high temperature. Heat is stored in LCZ, which should be sized to supply energy throughout the year. It is almost as thick (usually 1 m) as the NCZ. This is the heat collector, heat storage and heat removal medium. The bottom boundary is a black body.

Principle of a Solar Pond

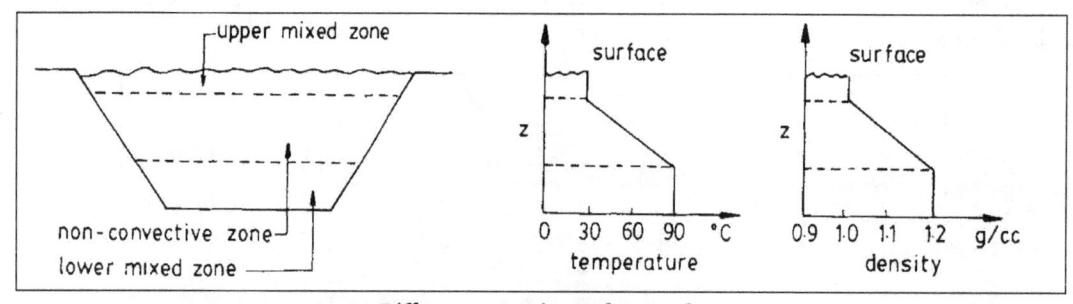

Different zones in a solar pond.

In a clear natural pond about 30% solar radiations reaches a depth of 2 metres. This solar radiation is absorbed at the bottom of the pond. The hotter water at the bottom becomes lighter and hence rises to the surface. Here it loses heat to the ambient air and, hence, a natural pond does not attain temperatures much above the ambient. If some mechanism can be devised to prevent the mixing between the upper and lower layers of a pond, then the temperatures of the lower layers will be higher than of the upper layers. This can be achieved in several ways. The simplest method is to make the lower layer denser than the upper layer by adding salt in the lower layers. The salt used is generally sodium chloride or magnesium chloride because of their low cost. Ponds using salts to stabilize the lower layers are called 'salinity gradient ponds'. There are other ways to prevent mixing between the upper and lower layers. One of them is the use of a transparent honeycomb structure which traps stagnant air and hence provides good transparency to solar radiation while cutting down heat loss from the pond. The honeycomb structure is made of transparent plastic material. One can also use a transparent polymer gel as a means of allowing solar radiation to enter the pond but cutting down the losses from the pond to the ambient.

A salinity gradient solar pond as this technology has made tremendous progress in the last fifteen years. Typical temperature and density profiles in a large salinity gradient solar pond are shown in figure. We find that there are three distinct zones in a solar pond. The lower mixed zone has the highest temperature and density and is the region where solar radiation is absorbed and stored. The upper mixed zone has the lowest temperature and density. This zone is mixed by surface winds, evaporation and nocturnal cooling. The intermediate zone is called the non-convective zone (or the gradient zone) because no convection occurs here. Temperature and density decrease from the bottom to the top in this layer, and it acts as a transparent insulator. It permits solar radiation to pass through but reduces the heat loss from the hot lower convective zone to the cold upper convective zone. Heat transfer through this zone is by conduction only. The thicknesses of the upper mixed layer, the non-convective layer and the lower mixed layer are usually around 0"5, 1 m and 1 m, respectively.

Thermal Performance

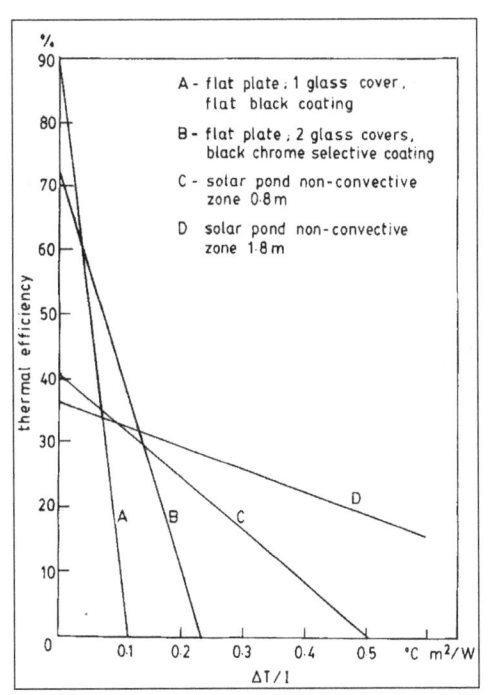

Variation of thermal efficiency with $\Delta T/l$, where ΔT is the temperature difference between the storage zone (or absorber) and the ambient, and I is the intensity of solar irradiation.

The thermal performance of a solar pond can be represented in a form similar to that used for conventional flat plate collectors. Assuming a steady state condition,

$$Q_u = Q_a - Q_e$$

Where, Q_u = useful heat extracted, Q_a = solar energy absorbed, Q_e = heat losses.

The thermal efficiency of a solar pond can be defined as $\eta = (Q_u / I)$ where I is the solar energy incident on the pond. Thermal effÉciency can be written as $\eta = \eta_o - Q_u/I$, where η_o is called the optical efficiency of the pond (Q_a/I). We express $Q_e = U_o(T_s - T_a)$, where T_s is the pond storage-zone temperature, T_a is the ambient temperature and U_o is the overall heat-loss coefficient. If we neglect heat losses from the bottom and sides of the pond and assume that the temperature of the upper mixed layer is the same as the ambient, then $U_o = K_w/b$ where, K_w is the thermal conductivity of water and b is the thickness of the gradient zone.

Kooi has compared the efficiency of a solar pond with those of conventional flat plate collectors. We find that the thermal efficiency of a solar pond is higher than that of a flat plate collector when the operating temperatures are higher, and is in the range of 20 to 30% when the temperature difference is around 60 °C. The thermal efficiency is strongly dependent upon the transparency of the pond which is influenced by the presence of algae and dust. Even if the solar pond is free of dust and algae, the absorption properties of pure water influence the transmission of solar radiation in the pond. The transmissivity of solar radiation in pure distilled water is shown in figure. We observe that about half the solar radiation is absorbed in the first 50 cm of water. This is on account of strong infrared absorption bands in water. At a depth of 2 metres the transmission is about 40%. This sets the upper limit on the thermal efficiency of a solar pond. The thickness of the gradient zone must be chosen depending on the temperature at which thermal energy is needed. If the thickness of the gradient zone is too high the transmission of solar radiation is reduced while if it is too small it causes high heat losses from the bottom to the top of the pond. The optimum value of the thickness depends on the temperature of the storage zone of the pond. Nielsen has provided a steady state analysis of a solar pond and has included the effect of solar radiation absorption in the gradient zone on the temperature profile. In the steady state, the energy equation becomes,

$$K(d^2T/dZ^2) = I(d\tau/dZ)$$

Where, K is the thermal conductivity of water and τ is the fraction of solar radiation I reaching a depth Z.

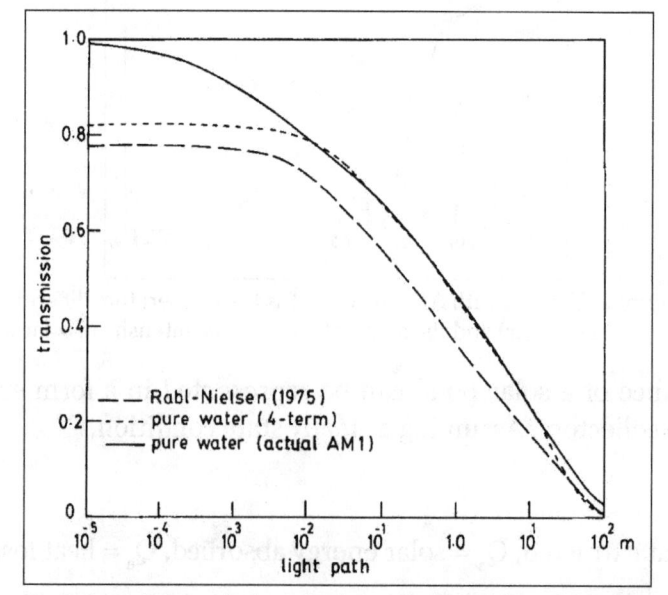

Variation of transmission of solar radiation with path length.

This equation can be integrated to get,

$$(dT/dZ) = \{I/K\}\{\tau(Z) - \tau(Z_1) + (dT/dZ)_{Z_2}\}$$

Where, Z_2 is the interface between the gradient zone and storage zone. If η is the fraction of the incident solar energy which is extracted from the system as heat (including ground losses), then an energy balance of the storage zone gives,

$$(dT/dZ)\big|_{Z_2} = \{I/K\}\{\tau(Z_2) - \eta\}$$

We can combine last two equations to obtain the temperature profile in the gradient zone as,

$$(dT/dZ) = \{I/K\}\{(\tau(Z) - \eta)\}$$

The temperature profile in the gradient zone for various values of η is shown in figure for I = 200 W/m². Since T is proportional to I, the above figure can also be used for other values of I by multiplying by an appropriate constant.

The effect of ground-heat losses on the performance of a solar pond has been analysed by Hull et al. They have shown that the ground heat-loss coefficient can be expressed as,

$$U_g = K(1/D + bP/A)$$

Where, K is the ground conductivity. D is the depth of the water table, P and A are the pond perimeter and surface area and b is a constant whose value is around 0.9 (depending upon the side slope). The thermal efficiency of a steady state solar pond can now be written as,

$$\eta = \frac{1}{Z_2 - Z_1}\left[\int_{Z_1}^{Z_2}\tau(Z)dZ - \frac{K_w \Delta T}{I}\right] - \frac{U_g \Delta T}{I}$$

Where, ΔT is the temperature difference between the storage zone and the upper mixed layer.

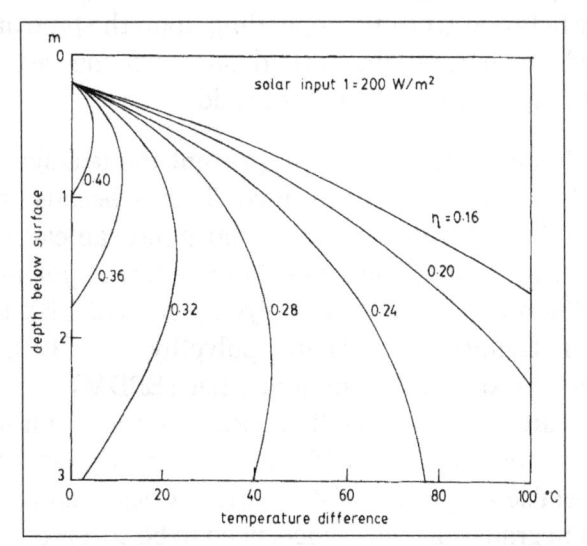

Variation of efficiency with depth and temperature difference.

Note that the optical efficiency of the solar pond is dependent upon the mean transmittance in the gradient zone. This is because the radiation absorbed in the gradient zone is helpful in reducing the heat losses from the storage zone.

The steady-state analysis of a solar pond is useful in the sizing of the pond for a specific application. There will, however, be strong seasonal variation in the performance of the pond on account of seasonal variations in solar insolation, wind and temperature. Srinivasan has proposed a simple two-zone model for the simulation of the storage zone temperature of the pond.

Pond Construction

The site selected for the construction of a solar pond should have the following attributes:

- Be close to the point where thermal energy from the pond will be utilized;
- Be close to a source of water for flushing the surface mixed-layer of the pond;
- The thermal conductivity of the soil should not be too high;
- The water table should not be too close to the surface.

An estimate of the area required for a solar pond (in the tropics) can be obtained from figure. To minimize heat losses and liner costs, the pond should be circular. Since a circular pond is difficult to construct, a square pond is normally preferred. For large solar ponds (area > 10,000m^2), the shape will not have a strong influence on cost or heat losses. The depth of the solar pond must be determined depending on the specific application. The usual thicknesses of the surface, gradient and storage zone of the pond are 0.5, 1 and 1 m, respectively. If a particular site has low winds, one can reduce the thickness of the surface layer to 30 cm. If the temperature required for process heat applications is around 40 °C (such as hatcheries) then the thickness of the gradient zone can be reduced to 0.5 m. Storage zone thickness higher than 1 m may be required to take care of long periods of cloudiness.

The excavation for a solar pond is similar to that for construction of water reservoirs. The side slope of the pond can vary between 1:1 to 1:3 depending upon the type of soil. After the excavation and bounding is completed, and before a liner is laid, one must ensure that the area is free of sharp objects which may damage the liner when it is being laid.

In most solar ponds, a polymeric liner is used to prevent the leakage of salt. Some solar ponds in Israel, Australia and Mexico have not been lined. This is because at those sites the soil has low permeability. Since the leakage of salt from a solar pond can cause environmental pollution it is necessary to use a liner in most applications. Many types of polymeric liner have been used in solar ponds. Some of them are low-density polyethylene (LDPE), high-density polyethylene (HDPE), plasticized polyvinyl chloride, chlorinated polyethylene (CPE), chloro-sulphonated polyethylene (Hypalon), ethylene propylene diene monomer (EPDM) and polymer-coated polyester fabric (XR-5). These liners are available usually in 10m widths and hundreds of metres lengths, and are heat-sealed in the factory or in the field. Liners such as XR-5, EPDM and Hypalon can be used as exposed liners because they can resist ultraviolet degradation. Liners such as LDPE or HDPE undergo ultraviolet degradation and, hence, need to be covered with soil, brickwork or tiles.

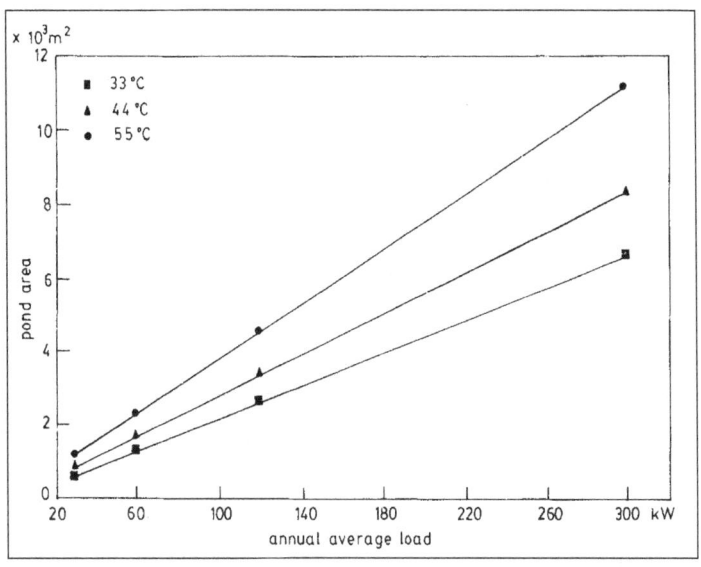

Solar pond area as a function of annual average pond load.

After the installation of the liner it can be tested for leaks by using a portable blower. If there are any pin holes or leakages at joints the liner will billow upwards. An inexpensive method to reduce leakage from pin holes is the use of Bentonite clay between adjacent LDPE liners. When Bentonite clay absorbs moisture it swells considerably and blocks further leakage. The use of alternate layers of LDPE and clay has been implemented at the 2,10,000 m² solar pond at Bet Ha Arava in Israel. After the installation of the liner, it is useful to have a method for detection of any leakage of salt. Hull et al have shown that an accurate calculation of salt inventory in the pond will provide indication of leak as low as 1 mm per month.

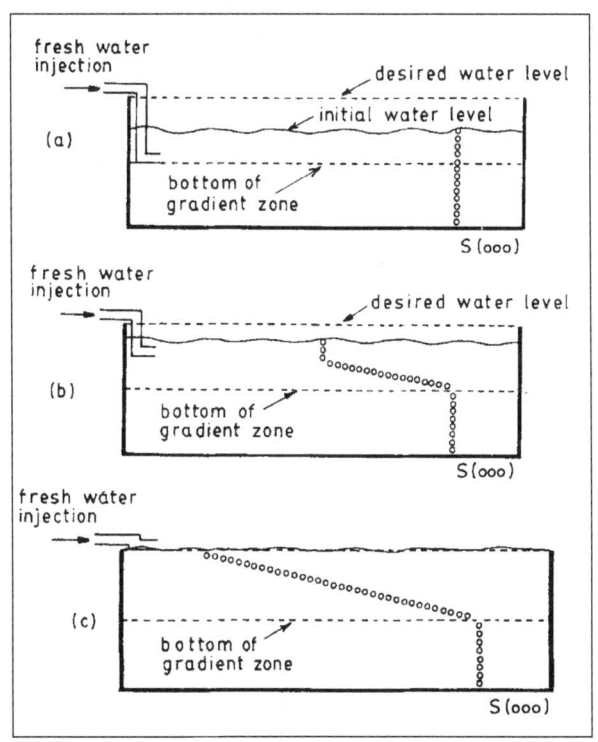

Establishment of initial (a), halfway (b) and final (e) gradient zones in a solar pond.

After the liner is placed, the pond is filled with water to a depth equal to the thickness of the storage zone and half the gradient zone. Salt is directly dumped into the pond. The salt dissolves rapidly if the water in the pond is circulated through a pump and the water is directed as a jet into the pond. The concentration of the salt at the storage zone is between 200 to 300 kg/m³. Hence the salt inventory is between 1/3 to 1/2 ton per m². The normal method of establishing the gradient zone is by injection of fresh water. This method, developed by Zangrando, is convenient and hence has been adopted in most solar pond installations in the world. Fresh water injection is initiated at the interface between the storage zone and the gradient zone using a diffuser. The fresh water rises to the top and reduces the density of the layer above the injection point. For every 1 cm rise in water level, the diffuser is lifted by 2 cm. When the diffuser is at the same level as the water surface the establishment of the gradient zone is completed. More fresh water is added above the gradient zone to create an upper mixed layer with a thickness of 30 to 50 cm. The evolution of density profile during the establishment of the gradient zone in the Miamisburg solar pond is shown in figure.

After the establishment of the gradient zone, the pond begins to heat up if clear sky conditions prevail.

Pond Stability

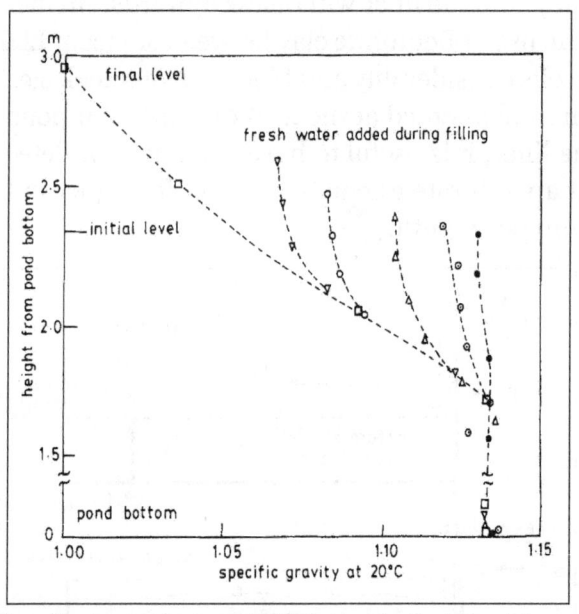

Evolution of gradient zone during establishment in the Miamisburg solar pond.

A solar pond will be statically stable if its density decreases with height from the bottom. A solar pond is subjected to various disturbances such as the wind blowing at the top surface and heating of the side walls. The criterion for dynamic stability of the pond is somewhat more stringent than that for static stability. This criterion can be obtained by perturbation analysis of the basic laws of conservation of mass, momentum and energy. The criterion for stability obtained from such an analysis can be written as,

$$\beta_T \frac{\partial T}{\partial Z} < \beta_S \frac{\partial S}{\partial Z} \left[\frac{Sc+1}{Pr+1} \right]$$

Solar Collector and Thermal Technologies

Where,

- $\beta_T = -\dfrac{1}{\rho}\left[\dfrac{\partial \rho}{\partial T}\right]$ = thermal expansion coefficient,
- $\beta_S = +\dfrac{1}{\rho}\left[\dfrac{\partial \rho}{\partial S}\right]$ = salinity expansion coefficient,
- Pr = Prandtl number,
- Sc = Schmidt number.

For typical conditions encountered in a solar pond, this result can be simplified to,

$$\frac{\partial S}{\partial Z} > 1.19 \frac{\partial T}{\partial Z}$$

Where, S is in kg/m³ and T in °C.

In order to prevent formation of internal convective zones within the gradient zone it is essential that the above criterion is satisfied at all points within the gradient zone. Hull et al have recommended that a safety margin of 2 is desirable.

Safety margin (SM) is defined as,

$$SM = \left(\frac{\partial S}{\partial Z}\right)_{actual} \bigg/ \left(\frac{\partial S}{\partial Z}\right)_{theoretical}$$

Where,

$$\left(\frac{\partial S}{\partial Z}\right)_{theoretical} = \frac{\beta_T}{\beta_S}\left[\frac{Pr+1}{Sc+1}\right]\frac{\partial T}{\partial Z}$$

In large solar ponds, side-wall heating would not be able to initiate the formation of internal convective zones and hence a safety margin around 2 should be adequate.

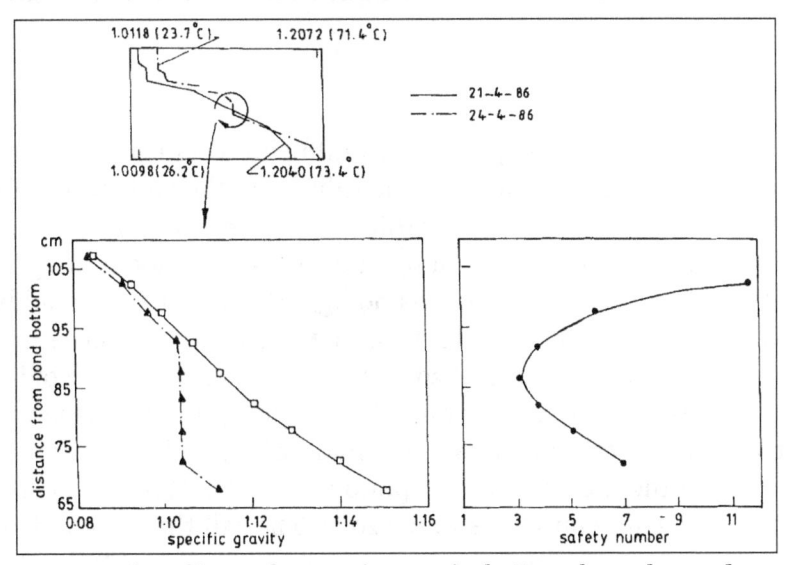

Formation of internal convective zone in the Bangalore solar pond.

The thickness of the gradient zone (which provides insulation and hence reduces heat losses) can be reduced by the formation of internal convective zones or erosion of the boundaries of the gradient zone. Erosion of the gradient zone-surface zone interface occurs primarily on account of wind-induced mixing. The effect of wind-induced mixing can be reduced by using floating plastic nets or pipes. Nielsen has reported that mean-squared wind speeds exceeding 20 m²/s² caused the erosion of the gradient zone at the rate of 1 cm/day, while for values below 10 m²/s² there was no gradient erosion.

Erosion of the gradient zone from below depends upon the density and temperature gradients at the gradient zone-storage zone interface. Nielsen has determined experimentally that the gradient zone-storage zone interface remains stationary if the salinity and temperature gradients satisfy the following relationship.

$$\frac{\partial S}{\partial Z} = A \left[\frac{\partial T}{\partial Z} \right]^{0.63}$$

Where,

$A = 28 \ (kg/m4) \ (m/K)0.63$

If the actual salinity gradient is more than that given by the above criterion, the gradient zone will move downwards, while if it is less, the gradient zone will erode.

To ensure that the gradient zone does not erode from above, the density of the surface layer must be kept as low as possible. The density of the surface layer increases on account of diffusion of salt from below and because of evaporation. Hence the surface layer must be flushed regularly with fresh water to keep the salinity below 5% (by weight).

Salt Replenishment

On account of the gradient of concentration between the storage and the surface zones, there is a steady diffusion of salt through the gradient zone. The transport of salt through the gradient zone by diffusion can be expressed as,

$Q_m = [(S_l - S_u) D] / b$,

Where, b = thickness of gradient zone, D = mass diffusion coefficient, and S_l, S_u = salinity in lower and upper mixed layers, respectively. If the salinity in the storage zone is 300 kg/m³ and in the surface zone is 20 kg/m³, gradient zone thickness is I m and diffusion coefficient of salt is 3×10^{-9} m²/s, then the rate of transport of salt by diffusion will be about 30kg/m 2 year. In small solar ponds the salt transport can be as high as 60 kg/m² year because of additional salt transport through side-wall heating. If the salt lost from the storage zone is not replenished regularly then there may be an erosion of the gradient zone from below or formation of internal convective zones. The normal method of salt replenishment is by pumping the brine in the storage zone through a salt bed; Srinivasan has shown that for small solar ponds a passive salt replenishment technique is adequate. In the Bangalore solar pond (240 m² bottom area) about 100 kg of salt was added daily through a chute into the storage zone. The salt that was added dissolved within a day.

Algae Control

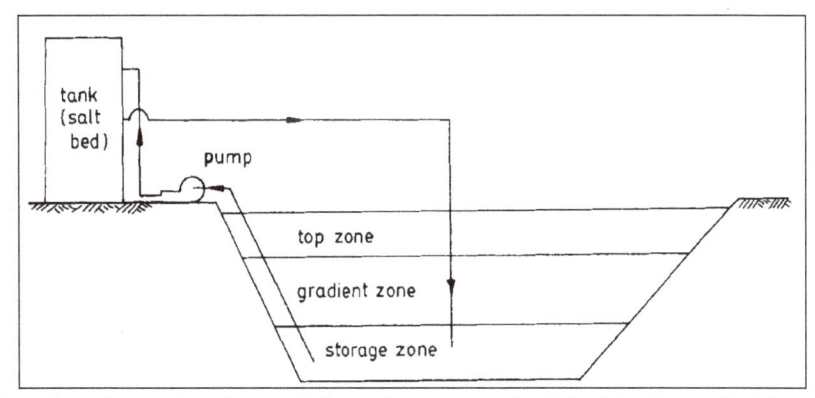

Salt replenishment in solar pond through an external salt bed (active replenishment).

The thermal efficiency of a solar pond is strongly dependent upon the clarity of the pond, which is reduced by the presence of algae or dust. Bits of debris/dust or leaves lighter than water float on the surface and can be skimmed off. Dust and debris much heavier than water will sink to the bottom. Srinivasan & Guha have reported that the dust accumulating at the bottom of the pond does not adversely affect the absorption of solar radiation at the bottom of the pond. The dust floating in the gradient zone can be settled by adding alum. The growth of algae can be controlled by adding bleaching powder or copper sulphate. If the water used in the pond is alkaline, copper sulphate will not dissolve. Hull has provided a detailed account of the relative merits of various methods of algae control.

Heat Extraction

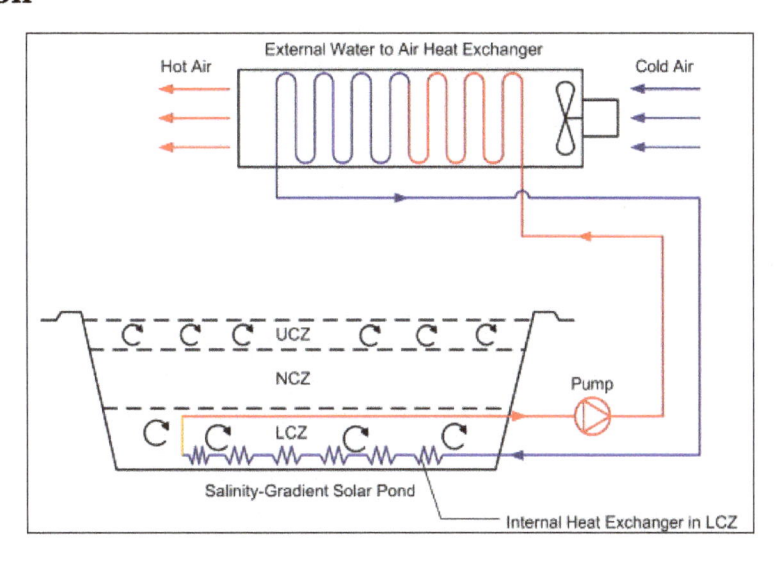

Heat can be extracted from the pond using an internal or an external heat exchanger. The internal heat exchanger is immersed in the storage zone. Such heat exchangers are made of copper or plastic to eliminate the effect of corrosion, and are appropriate for small solar ponds. In large solar ponds (area > 1000 m²) external heat exchangers may be more convenient. These are made of stainless steel or titanium.

References

- Linear-fresnel-reflector, the-4-types-of-csp-electricity-technologies, techologies-plants: estelasolar.org, Retrieved 15, April 2020
- Solar-collector-3006: corrosionpedia.com, Retrieved 26, May 2020
- Parabolic-trough-collector, engineering: sciencedirect.com, Retrieved 30, June 2020
- Linear-fresnel-systems-and-the-future-for-concentrated-solar-power: helioscsp.com, Retrieved 26, August 2020
- Solar-pond-and-its-application-to-desalination-290394062: researchgate.net, Retrieved 16, February 2020

Chapter 3

Principles of Solar Energy Generation

In order for the solar radiations to be used, it has to be captured, stored and then converted to electricity or any other form of power for consumption. The solar radiation can be harnessed in two different ways. Solar light can be used in power cooking appliances, lamps, panels or batteries. Solar heat can be used for heat generation and powering solar thermal stations to produce electricity. This chapter provides a detailed study of solar energy generation and its principles.

The energy from the sun amounts to 4x1020 MW, of which Earth receives only less than 1 % of the energy. This energy received from the sun can be harnessed directly or indirectly using various technologies for thermal applications as well as for converting into electricity by the means of photovoltaic (PV) systems. Over the years the photovoltaic technology advanced a lot and the efficiency of solar cell has considerably improved. As majority of our energy requirements are in the form of electricity, PV works on the principle of photovoltaic effect. The generation of thermal energy from solar can be realized using various solar reflecting collectors. Most of the technology works on the principle of reflection, radiation and convention or based on the thermosiphon effect.

Solar power generation technologies are important for providing a major share of the clean and renewable energy needed in the future, because they are the cost-effective among renewable power generation technologies. Solar power generation becomes sustainable and competitive with fossil-fuel power generation within the next decade. Solar power generation has proven to be one most attractive option for electrical energy production in grid-connected and distributed modes. The solar power generation can be done both by photovoltaic (PV) and concentrating solar power (CSP) systems. The PV technologies such as single and multi-crystalline, thin film cells, organic/inorganic dye-sensitized, and multi-junction solar cells have seen an increasing trend for their utilities as backup energy generation systems for small-scale and rooftop applications. The CSP technologies such as solar dish, parabolic trough, linear Fresnel reflectors, and power tower are gaining momentum for large-scale solar power generation using power cycles/engines. This special issue on solar power generation is focused mainly various technologies, materials, and control strategies for effective solar energy conversion, energy storage, control, and implementation approaches.

Various novel solar PV systems are proposed for performance improvement and cost effectiveness. Development of materials plays major role in viable solar power generation.

Energy from Solar Spectrum

Sun is a gigantic star, with diameter of 1.4 million kilometer releasing electromagnetic energy of about 3.8×10^{20} MW. The energy from the sunlight extends from 300nm to 3000 nm. Majorly, they are classified as Ultraviolet region (less than 350nm), Visible region (350 nm to 750 nm), and Infrared region (more than 750 nm). These various components of the sunlight constitute the solar spectrum. The visible (47 %) and infrared (46 %) components of the solar radiation contributes for most of the solar energy. It is important to understand, in general, the spectrum of the sun energy, as the technology used for energy generation and conversion is driven by the inputs received from the respective spectrum of solar irradiance.

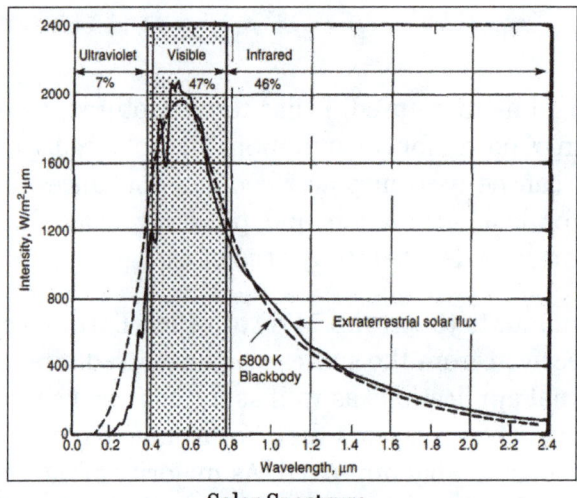

Solar Spectrum.

The energy from heat and light of solar radiation can be extracted to useful applications and the principle of operation is different depending on the technology. The PV technology convert visible spectrum to electricity and thermal collectors use both infrared and visible spectrum for energy generation.

So the energy generation from solar radiation can be in the form of electrical energy or thermal Energy. The various conversion paths of solar energy is described in the figure.

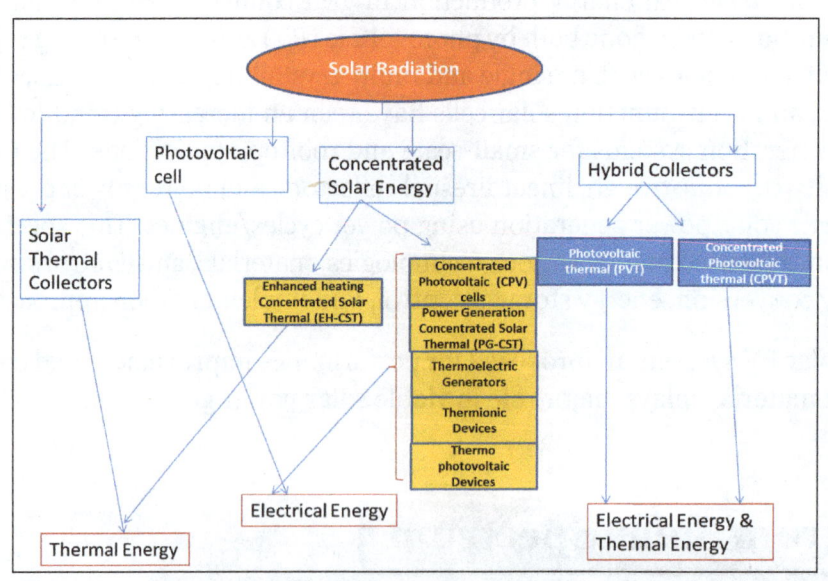

Various conversion paths of Solar Energy into electrical and thermal energy.

Principle of Electricity Generation by Solar Photovoltaic

The solar photovoltaic works on the principle of photovoltaic effect. It is the physical and chemical property or phenomenon in which electromotive force is generated in the non-homogeneous materials with the illumination of light of a specific wave length. This effect produces voltage and electric current in a material upon exposure to light. The photovoltaic property is seen in semiconductors when photons or radiant energy falls on surface is capable of converting to electric current.

There are majorly three energy bands in a semiconductor material:

- Filled band,
- Conduction band,
- Forbidden band.

Valance Band

The electrons move in the atoms in certain energy levels but the energy of the electrons in the innermost shell is higher than the outermost shell electrons. The electrons that are present in the outermost shell are called as Valance Electrons.

These valance electrons, containing a series of energy levels, form an energy band which is called as Valence Band. The valence band is the band having the highest occupied energy.

Conduction Band

The valence electrons are so loosely attached to the nucleus that even at room temperature; few of the valence electrons leave the band to be free. These are called as free electrons as they tend to move towards the neighboring atoms.

These free electrons are the ones which conduct the current in a conductor and hence called as Conduction Electrons. The band which contains conduction electrons is called as Conduction Band. The conduction band is the band having the lowest occupied energy.

Forbidden Gap

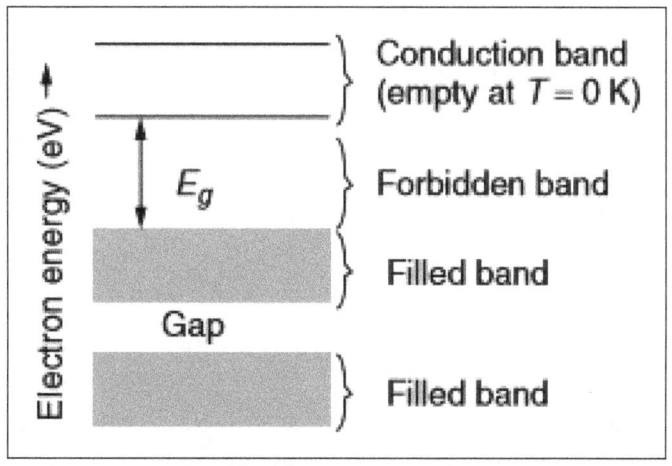

Energy bands in a semiconductor.

The gap between valence band and conduction band is called as forbidden energy gap. As the name implies, this band is the forbidden one without energy. Hence no electron stays in this band. The valence electrons, while going to the conduction band, pass through this.

The forbidden energy gap if greater means that the valence band electrons are tightly bound to the nucleus. Now, in order to push the electrons out of the valence band, some external energy is required, which would be equal to the forbidden energy gap.

The forbidden band is so wide, that an electron has to acquire sufficient energy to move from filled band to conduction band. This energy is known as band gap energy. The band gap energy in a silicon semiconductor is near to 1.12eV. The energy bands for other semiconductors like Gallium arsenide is 1.42eV, and Cadmium Telluride is 1.5eV.

The energy gap of 1.12eV indicates that, the photons with energy in the wavelength corresponding to 1.12eV can initiate the electrons to jump and create current flow. Any wavelength, below to that, adds to heat losses in the cell. In a p-n junction as long as the energy from the sun strikes, hole-electron pair is created and continuously recombined. If the recombination is avoided, and electron is forced to flow in one direction will facilitate the current flow in the system. Thus is made possible by a built in electric filed created within the semiconductor by making the semiconductor, impure/extrinsic. An impure semiconductor will have an additional electron or an additional hole in it. Usually, a pentavalent impurity is added to create excess electrons, and a trivalent impurity addition creates an excess hole.

Calculation of Band Gap Energy

Example: Find out the band gap energy for a semiconductor transparent to light of wavelength 0.87 μm?

Solution: Photon energy of light with 0.87 m wavelength.

Speed of light = c (3 x 10 8 m/s).

$$h\nu = c/\lambda$$

$$= \frac{6.63 \times 10^{-34} (J.s) \times 3 \times 10^8 \, m/s}{0.87 \, \mu m} = \frac{1.99 \times 10^{-29} (J. \mu m)}{0.87 \, \mu m}$$

$$= \frac{1.99 \times 10^{-19} (eV \, \mu m)}{1.6 \times 10^{-19} \times 0.87 \, \mu m} = \frac{1.24 (eV. \mu m)}{0.87 \, \mu m}$$

$$= 1.42 \, eV$$

So the band gap energy of semiconductor is 1.42 eV.

The relationship can be simplified as,

$$h\nu \, (eV) = 1.24 / \lambda \, (\mu m)$$

The solar cell is a p-n junction with large surface area. The n-type material is thin for passing light through it and strike the p-n junction. The electricity is generated inside the depletion zone of the p-n junction. When a photon of light is absorbed by one of the atoms in n-region of silicon, it dislodge an electron from any atom creating a free electron and hole pair. This free electron

Principles of Solar Energy Generation 115

and hole pair has sufficient energy to jump out of depletion zone. If a wire is connected from the cathode at n-type silicon to an anode of p-type silicon, electrons flow through the wire. The electron is attracted to the positive charge of p-type material and travels through the external load creating flow of electric current. The vicinity of a p-n junction when exposed to light is shown in the figure.

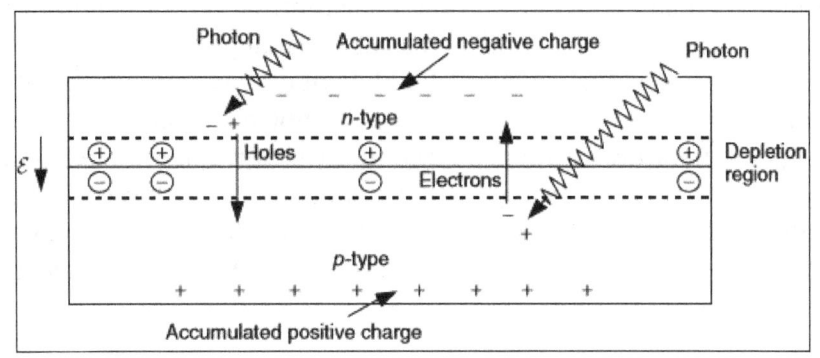

Vicinity of p-n junction.

The current generated from the solar cell can be represented to an equivalent p-n junction circuit response. An equivalent circuit for p-n junction solar cell is given in the figure.

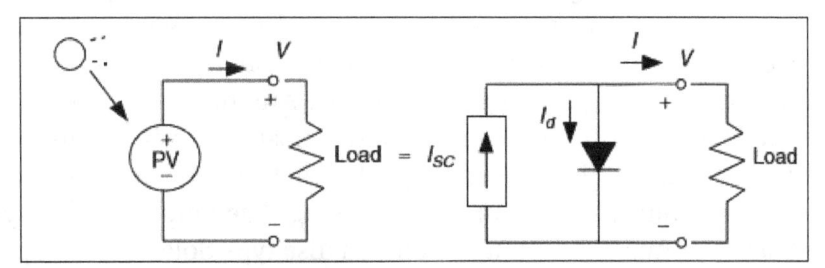

Equivalent circuit for p-n junction solar cell.

The intensity of the incident radiation and external load of the cell determines I-V characteristics of a solar cell. The voltage and current generation from the solar cell can be easily calculated from the equivalent circuit.

Factors Affecting the Energy Generation in a Solar PV cell Technology

The two main parameters which affect the performance output of a PV cell are temperature and the light (photons) incident on it. The voltage output is driven by the change in the temperature, and the current output is driven by the light received. The increase in the light input, contributes to increase in the current output. However, the junction voltage reduces with the increase in the temperature.

Solar Thermal Power Generation

Solar Power Generation Technology

Solar power generation technology is mainly divided into two types, one is solar light power generation technology, and the other is solar Solar-thermal power generation technology. Solar power generation mainly includes photovoltaic power generation, photochemical power generation, optical induction power generation and biological power generation, among which photovoltaic power generation technology is widely used. Photovoltaic power generation has the characteristics of high efficiency, low pollution and good flexibility, but photovoltaic panels have many defects such as high pollution, high energy consumption and large space occupation. Solar thermal power generation technology mainly includes tower solar thermal power generation system, trough solar thermal power generation system and dish solar thermal power generation system. Than solar-thermal power generation is the sun point-blank light energy through the adoption of many a mirror together, make the heat transfer fluid in the heat pipe heat continuously heating up, and to transfer heat to high temperature steam produced in the steam generator, and then by the power generating units of a renewable energy application technology, it has low cost, stable output power and power generation system of continuous adjustable, and long service life, no pollution and other advantages, therefore, solar-thermal power generation is more competitive power generation technology.

Technical Analysis of Solar Thermal Power Generation

According to the different power generation principles, Solar-thermal power generation includes concentrated Solar-thermal power generation, solar semiconductor temperature difference power generation, solar chimney power generation, solar pool power generation and solar thermal acoustic power generation. Among them, concentrated Solar-thermal power generation is the most commercial use of the most promising technology. Among them, concentrated Solar-thermal power generation is the most commercial use of the most promising technology. According to the different ways of condensing, the condensing Solar-thermal power generation can be further divided into two systems: point focusing and line focusing. The point focusing system mainly includes tower type Solar-thermal power generation and disc type Solar-thermal power generation. The line-focusing system mainly includes trough Solar-thermal power generation and linear Fresnel Solar-thermal power generation.

Principle of Solar Thermal Power Generation

Solar-thermal power generation principle is that through the reflectors, such as condenser of heat exchanger will collect solar radiation into heat energy collection of hot charging, used to heat the heating device inside the heat transfer medium, such as heat conduction oil or molten salt with a heat exchange device, heat transfer medium water heated to high temperature and high pressure steam, steam to drive a turbine driven generator to produce electricity. This through the "light - heat - mechanical - electrical energy transformation process of the realization of power generation technology is known as concentrated solar power technology. The principle and basic equipment composition of solar thermal power generation are basically the same as those of fossil fuel power plants. The biggest difference is that the heat sources used for power generation are different. Solar thermal power generation USES clean and abundant solar energy.

Solar Thermal Power Generation Technology Types

Tower Solar Thermal Power Generation System

Tower type solar thermal power generation is also known as concentrated solar thermal power generation. It takes the form of a number of arrays of mirrors that reflect solar radiation onto a solar receiver located at the top of the tower, heating the working medium to produce superheated steam, which drives a turbine generator to generate electricity and convert the absorbed solar energy into electricity. Tower solar thermal power generation is mainly composed of four parts: mirror field, heat exchange system, heat storage device and steam turbine generator. Tower solar thermal power generation system is shown in figure.

Tower solar thermal power generation system.

The main features of the tower solar thermal power generation system are as follows: (1) the concentration-light ratio usually achieved by the tower solar thermal power generation system is 300 ~ 1,500, and the operating temperature can reach 1,000 ~ 1,500 °C.(2) the tower Solar-thermal power generation system has short heat transmission distance, low heat loss and high comprehensive efficiency, which can reach about 14% at present; (3) solar tower power generation is suitable for large-scale and large-capacity commercial application; (4) the tower Solar-thermal power generation system has large one-time investment, complex device structure and control system, and high cost.

Trough Solar Thermal Power Generation System

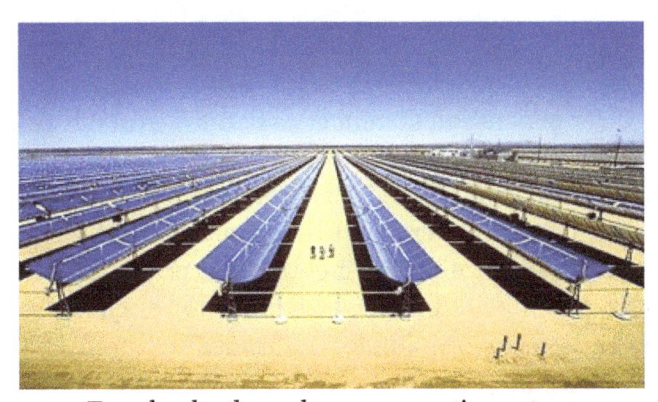

Trough solar thermal power generation system.

Trough type solar thermal power generation system is to use the groove parabolic mirror concentrated solar thermal power generation system. The focusing mirror from the point of view of geometry is the parabola translation and formation of the parabolic trough type, it will be the sunlight in a line, the focal online installation has tubular collector, after focusing to absorb solar radiation energy, and often many groove parabolic series-parallel into concentrating collector array. Slot to track one dimensional parabolic face more solar radiation (axis equipment placed between the north and the south, then what the track).Its geometric concentration ratio is between 10 and 100, and the temperature can reach about 400 °C. The trough solar thermal power generation system is shown in figure. At present, trough power station has the lowest operation risk and generation cost, and the most commercial value, which is suitable for medium-low temperature solar thermal power generation system.

Disc Solar Thermal Power Generation System

Disc type solar thermal power generation system using disk parabolic mirror to focus the sun's rays, installed in the focus of working medium heat absorber absorbs solar radiation heat absorption of heat, heat absorption working medium and working medium of the steam generator heat exchange water, generated by the high temperature and high pressure steam driving turbine generator, disc type solar thermal power generation system as shown in figure. The advantages of the system are that the concentrator ratio can reach 3,000, the receiver's heat absorption area is small, the working medium's heat collection temperature is > 800 °C, and the system efficiency can reach 29.4% at most.

Disc solar thermal power generation system.

Linear Fresnel Type Solar Thermal Power Generation System

Linear Fresnel thermal power generation system is similar to parabolic trough thermal power generation. It consists of many horizontal mirrors rotating on a single axis, which form a rectangular mirror to automatically track the sun. The reflected sunlight is gathered on the collector tube, and the fluid medium in the heating tube generates steam directly or indirectly, which drives the steam turbine unit to generate electricity. Linear Fresnel thermal power generation system is relatively simple, and the mirror can adopt plate mirror, which has lower cost but lower system efficiency. The structure of the system is relatively simple, the transmission mechanism is easy to operate, and the collector pipe can be made of steel. Therefore, the cost is lower than the tank system, so it can be applied in many places with its own characteristics, such as heating water/steam, providing steam to buildings and factories (temperature range: 80 °C ~ 250 °C), using small and medium-sized heat engines for medium-low temperature power generation, heating, refrigeration and other multi-generation, solar desalination and so on. However, the focus of this heat collection system is relatively small, so the temperature rise is limited, the heat collection tube needs to absorb heat and dissipate heat at the same time, so the heat loss in operation is relatively large, and the system efficiency is lower than that of the tank system.

Comparative Analysis of Solar Thermal Power Generation Technology

The characteristics of the above four solar photovoltaic power generation technologies are compared and analyzed, and the results are shown in table.

Project	Tower	Trough	Disc	Fresnel
Heat transfer medium	Water/steam, molten salt	Water/steam, molten salt, heat conducting oil	Molten salt	Water/steam
Focusing technology	Point focusing	Line focusing	Point focusing	Line focusing
Scale (MW)	30-100	30-350	5~25	10-320
Energy storage	Yes	Yes	No	Yes
Application	Grid-connected power generation	Grid-connected power Generation	Small capacity decentralized power generation, remote areas independent system power supply	Small application range
Concentrated ratio	300-1500	50-100	600-3000	25-150
Unit efficiency	23%	21%	30%	20%

As can be seen from table, trough Solar-thermal power generation has the most extensive application range, up to 30-350mw, and is suitable for middle and low temperature heating, with a generation efficiency of up to 21%. The concentration-light ratio of the tower solar thermal power system can reach 300 ~ 1,500, and the operating temperature can reach 1,000 ~ 1,500 °C. The tower Solar-thermal power generation system can be connected to the grid, and the power generation efficiency can reach about 23% at present.

However, the performance, initial investment and operation cost of the tower Solar-thermal power generation system are not sufficiently commercialized, and the tower power generation cost is high. The concentrator ratio of disc Solar-thermal power generation is 600-3000, the receiver's heat absorption area is small, the working medium's heat collection temperature is > 800 °C, and the system efficiency is up to 30%. However, the structure of disc solar collector is relatively complex, and the reliability needs to be strengthened. Fresnel Solar-thermal power generation system is characterized by simple system, direct use of thermal conductivity to generate steam, its construction and maintenance cost is relatively low. However, its focus is relatively small, the temperature rise is limited, and the heat loss during operation is relatively large. Overall, trough solar thermal power generation system is the most mature in technology, easy to realize, the overall cost is the lowest, and the heat collection temperature is moderate, more suitable for low and medium temperature solar thermal power generation system.

Thermodynamic Cycles for Solar Thermal Power Generation

The thermodynamic cycles used for solar thermal power generation can be broadly classified as low, medium and high temperature cycles. Low temperature cycles work at maximum temperatures of about 100 °C, medium temperature cycles work at maximum temperatures up to 400 °C, while high temperature cycles work at temperatures above 400 °C. Low temperature systems use fiat-plate collectors or solar ponds for collecting solar energy. Recently, systems working on the solar chimney concept have been suggested. Medium temperature systems use the line focussing parabolic collector technology. High temperature systems use either paraboloidal dish collectors or central receivers located at the top of towers.

Low Temperature Systems

A diagram of a typical low temperature system using flat-plate collectors and working on a Rankine cycle is shown in figure. The energy of the sun is collected by water flowing through the array of fiat-plate collectors. In order to get the maximum possible temperature, booster mirrors which reflect radiation on to the fiat-plate collectors are sometimes used. The hot water at-temperatures close to 100°C is stored in a well-insulated thermal storage tank. From here it flows through a vapour generator through which the working fluid of the Rankine cycle is also passed. The working fluid has a low boiling point. Consequently, vapour at about 90 °C and a pressure of a few atmospheres leaves the vapour generator. This vapour then executes a regular Rankine cycle by flowing through a prime mover, a condenser and a liquid pump. The working fluids normally used are organic fluids like methyl chloride and toluene, and refrigerants.

It has to be noted that the overall efficiency of this system is rather low, because the temperature difference between the vapour leaving the generator and the condensed liquid leaving the condenser is small. For the cycle shown in figure, the temperature difference is only 55 °C. This leads to a Rankine cycle efficiency of 7 to 8%. The efficiency of the collector system is of the order of 25%. Hence an overall efficiency of only about 2% is obtained.

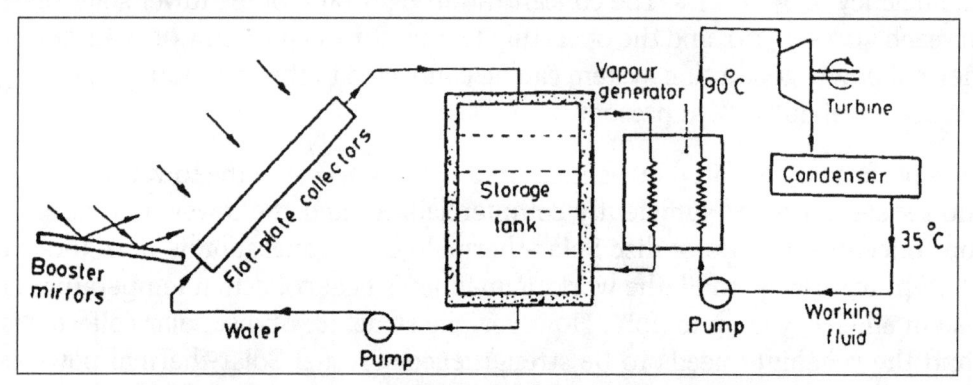

Low temperature power generation cycles using flat-plate collectors.

Plants of this type of French design having generation capacities up to about 50 kW were installed in many parts of the world.

In order to reduce the cost, solar ponds have been used instead of flat-plate collectors. The first two solar pond power plants having capacities of 6 kWe and 150 kWe were constructed in Israel about 15 years ago. These were followed in 1984 by the Bet Ha-Arava power plant, the largest in the world with an area of 250000 m^2 and a capacity of 5 MWe. The working of these plants has firmly established the technical viability of solar pond power plants. However they also do not appear to be economically attractive inspite of being less costly than plants using flat-plate collector systems.

Recently the concept of a solar chimney power plant has been suggested. In such a plant, a tall central chimney is surrounded at its base by a circular green-house consisting of a transparent cover supported a few metres above the ground by a metal frame. Sunlight passing through the transparent cover causes the air trapped in the green-house to heat up. A convection system is set up in which this air is drawn up through the central chimney turning a trubine located near the base of the chimney. The hot air is continuously replenished by flesh air drawn in at the periphery of the green-house.

The only solar chimney power plant built so far is a 50 kW pilot plant in Spain. It has a 200 m high chimney with a constant diameter of 10.3 m. The solar collector area extends to a radius of 126m from the chimney with the glazing being 2m above the ground. The turbine, housed at the base of the chimney, has four 5 m long blades and rotates at 1500rpm to produce an output of 50kW.

Although the energy conversion efficiency of such plants is low (of the order of 1%), it is claimed that there will be considerable reduction in cost with scale-up and that a large size 1000 MW plant may cost only $1000 per kW.

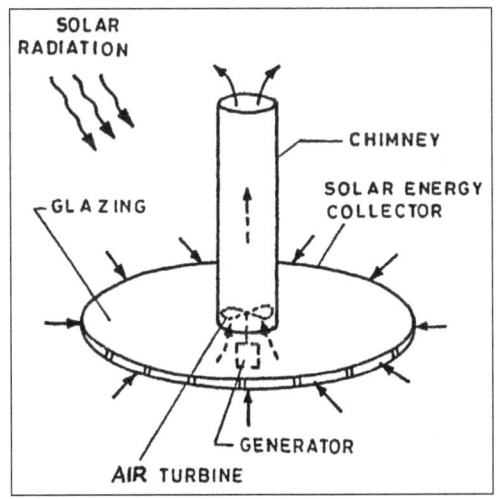

Solar chimney power plant.

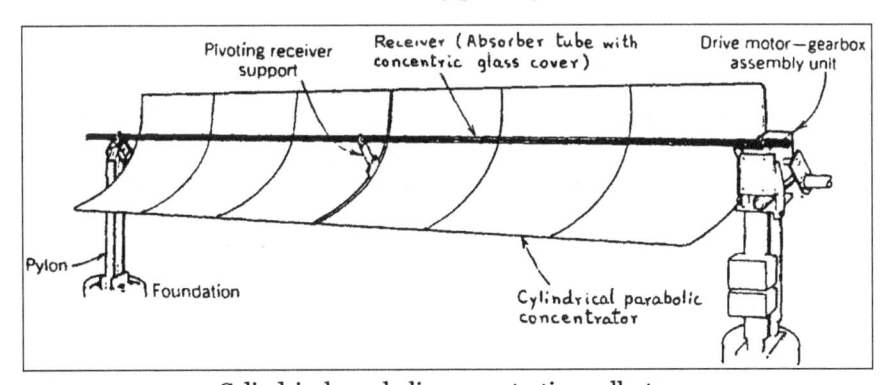

Cylindrical parabolic concentrating collector.

Medium Temperature Systems

Among solar thermal-electric power plants, those operating on medium temperature cycles and using line focusing parabolic collectors at a temperature of about 400 °C have proved to be the most cost effective and successful so far. A schematic diagram of a typical plant is shown in figure. The first commercial plant of this type having a capacity of 14 MW was set up in 1984. Since then, six plants of 30 MW capacities each, followed by two plants of 80 MW each have been installed and commissioned. All these plants have been set up by LUZ International in California, which has a total installed capacity of 354 MW. The collector array for the 80 MW plant has an area of 464340 m². The cylindrical parabolic collectors used have their axes oriented north-south. The absorber tube used is made of steel and has a specially developed selective surface. It is surrounded by a glass cover with a vacuum. The collectors heat synthetic oil to a temperature of 400 °C with a collector efficiency of about 0·7 for beam radiation. The synthetic oil is used for generating super-heated high pressure steam which executes a Rankine Cycle with an efficiency of 38%. The plant generally produces electricity for about 8 h a day and is coupled with natural gas for continuous operation. The installed cost of this type of plant has reduced over the years because of the increasing installed capacity. The latest 80 MW plant is reported to have cost $.2900 per kW. The current generating cost is about 8 cents per kWh.

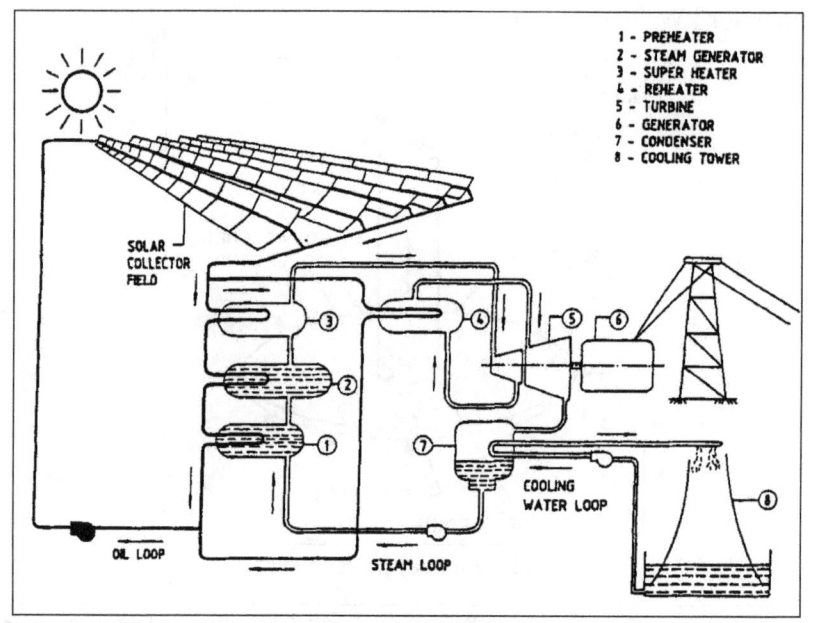

Medium temperature power generation cycle using parabolic concentrating collectors.

The following are some details of the collector modules used in the 80 MW power plant.

Aperture	5.76m
Length	95.2m
Reflecting surface	224 curved mirror glass panels
Reflectivity	0.94
Glass cover transmissivity	0.965
Vacuum in annular space	10^{-4} torr
Absorber tube O. D.	0.070 m
Tube surface absorptivity	0.97
Tube surface emissivity	0·15
Optical efficiency	0.772
Peak collection efficiency	0.68 (based on beam radiation)
Annual collection efficiency	0.53 (based on beam radiation)

The above data give some idea of the international state-of-the-art in line focussing cylindrical parabolic collector technology. It may be worth noting that such collectors are not yet being made commercially in India. However considerable expertise has been developed in constructing a number of prototypes in many research institutions.

High Temperature Systems

Two concepts have been tried with high temperature systems. These are the paraboloidal dish concept and the central receiver concept.

Paraboloidal Dish Collector System

In the paraboloidal dish concept the concentrator tracks the sun by rotating about two axes and the sun's rays are brought to a point focus. A fluid flowing through a receiver at the focus is heated and this heat used to drive a prime mover. Typically Stirling engines have been favoured as the prime movers and systems having efficiencies upto 30% and generating power in the range of 8 to 50 kW have been developed. The Indian experience with this type of system has been restricted to a small experimental 20 kW power station near Hyderabad. Four paraboloidal dish collector modules were used to generate steam which ran a steam engine. Becuase of limitations on the size of the concentrator, paraboloidal dish systems can be expected to generate power in kilowatts rather than megawatts. Thus they can be expected to meet the local power needs of Communities, particularly in rural areas.

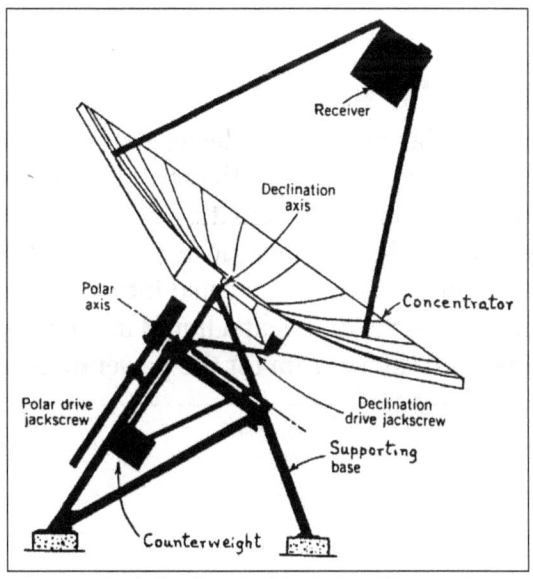

Paraboloid concentrating collector.

Some commercial designs of paraboloid dish collector systems have been developed abroad in the last ten years for electric power production.

A 7.5 m diameter stretched metal membrane concentrator has been developed by a German firm. The membrane is a stainless steel sheet (0.23 mm thick) fixed on both sides of a circular ring. The two membranes are deformed plastically to a parabolic shape by applying a water load and a partial vacuum, the vacuum being maintained during operation of the concentrator. The front membrane is covered with thin glass mirrors having a reflectivity of 0.90 and an area of 42 m^2. The concentrator is suspended at two points in a polar mounting and tracks the sun by rotating daily about a vertical axis and seasonally about a horizontal axis. The focal length is 5 m. A cavity-type receiver having a diameter of 0.2 m is kept at the focus. About 27 kW of energy is absorbed in the receiver if the incident beam radiation is 800 W/m². A Stirling engine located at the focus converts this thermal input to 8 kWe with an energy conversion efficiency of 0.3. More recently, the same firm has built two 17 m diameter dishes of the same design generating 50 kW each. These are in operation in Saudi Arabia. Dish/Stirling engine systems have also been built by other manufacturers.

It is generally felt that paraboloid dish systems are best suited for applications which utilize solar energy directly at the focus of each collector. However in USA, a 5 MW power plant utilizing the steam generated by seven hundred dishes has been erected. Each dish consists of a reflecting array of twentyfour 1.5 m diameter mirrors having an area of 42 m². The mirrors are made of reflective polymeric film fixed on circular aluminium frames and subjected to a continuously applied vacuum. The receiver is an insulated cylindrical cavity about 0.9 m long and 0.6m diameter and contains a molten salt. Pipes carrying water/steam pass through the salt bath. Thus the solar energy is first absorbed by the molten salt and then transferred to the water/steam, the salt bath acting as a storage which takes care of small variations in solar radiation.

Out of the total number of seven hundred dishes, six hundred are used to obtain saturated steam at 275 °C, while the remaining one hundred dishes are used to superheat the steam to 400 °C. The steam is used to run two turbine-generator sets-one a main set of 3.68 MW and the other, a peaking set of 1-24 MW.

Central Receiver Power Plant

In central receiver power plants; solar radiation reflected from arrays of large mirrors (called heliostats) is concentrated on a receiver situated at the top of a supporting tower. A fluid flowing through the receiver absorbs the concentrated radiation and transports it to the ground where it is used to operate a Rankine power cycle. A schematic diagram showing the main components of a central receiver power plant in which water is converted into steam in the receiver itself is shown in figure. Alternatively, the receiver is used to heat a liquid metal or a molten salt and this fluid is passed through a heat exchanger in which steam for the power cycle is generated.

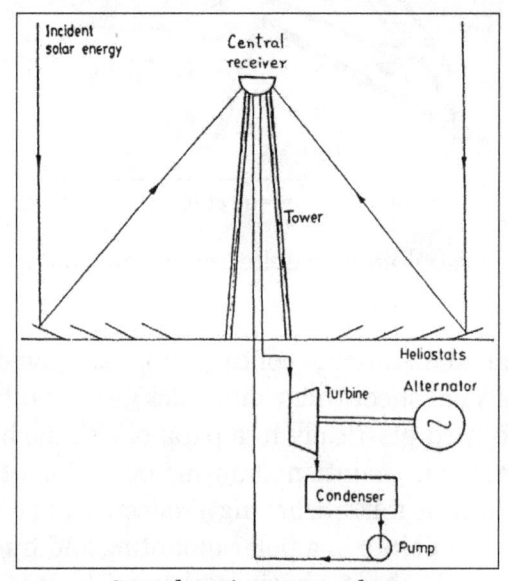

Central receiver power plant.

The idea of building such a plant was first suggested by scientists in the Soviet Union. Based on their calculations, they indicated the possibility of erecting an installation in the sunny regions of the USSR to produce 11 to 13 t of steam per hour at 30 atm and 400 °C. The optical system was calculated to consist of 1293 mirrors of 3 × 5 m. These heliostats were proposed to be mounted on carriages which moved on rails in arcs around the tower.

A number of small pilot plants were built in Italy in the period 1965 to 1967. In one of these 50 kW of energy was collected. After a break of a few years, the design of central receiver collector systems again attracted attention in the eighties and seven plants ranging in capacity from 0.5 to 10 MWe were built. These are listed in table along with some technical specifications. These include the number and the size of the heliostats, the receiver type, the receiver fluid and the height of the central supporting tower.

PLANT NAME	SSPS	EURELIOS	CESA I	SUNSHIN	THEMIS	CES	SOLAR ONE
Location	Spain	Italy	Spain	Japan	France	USSR	USA
Output (MWe)	0.5	1	1.2	1	2	5	10
Number of heliostats	93	112.70	300	807	201	1600	1818
Area of heliostat(m^2)	39.30	23.52	39.60	16	53.70	25	39.30
Total reflecting area (m^2)	3655	6216	11880	12912	10740	40000	71447
Receiver type	Cavity	Cavity	Cavity	Cavity	Cavity	External	External
Receiver fluid	Sodium	Steam	Steam	Steam	Molten Sait	Steam	Steam
Tower height (m)	43	55	60	69	-	70	80
Start of operation	1981	1981	1983	1981	1983	1985	1982

Although all the central receiver plants have been operated successfully, the available data indicate that the construction cost was very high. For example, the largest plant, Solar One, at Barstow, California cost approximately $14, 000 per kW. However, costs are likely to reduce with more operational experience, improved design and scale-up.

The two major components requiring considerable development are the heliostats and the receiver. These will now be discussed.

The heliostats form an array of circular arcs around the central tower. They intercept, reflect and concentrate the solar radiation onto the receiver. The array is served by a tracking control system which continuously focuses beam radiation towards the receiver during collection. In addition, when solar radiation is not being collected, the control system orients the heliostats in a safe direction so that the receiver is not damaged.

As stated earlier, the 10 MWe plant at Barstow was the largest of the pilot plants built. It was operated for six years from 1982 to 1988. The plant had a field of 1818 heliostats positioned all round a central tower of height 80 m. Each heliostat was an assembly of 12 slightly concave glass mirrors mounted on a support structure and geared drive that could be controlled for azimuth as well as elevation. The total reflective area of each heliostat was 39.3 m^2. A rear view sketch is shown in figure. The 12 mirror panels in each heliostat were 1 x 3 m in size and were made from 3 mm low iron float glass. When clean, the heliostats had an average reflectivity of 0-903. However exposure to the environment caused them to become dirty, thereby reducing the average reflectivity to 0.82. A goal of 0.92 has been set for future heliostat arrays. In order to achieve this goal along with reduced cost and weight, a. number of new concepts are being tried. For example, larger size glass-mirror heliostats having areas of 150 m² and reflectivity values up to 0-94 have been built. Also a new type of cost effective heliostat using a stretched membrane has been developed. In this heliostat, the reflector is a silvered polymer film laminated to a thin metal foil which is stretched over a large-diameter metal ring. The reflectivity of this surface has been measured to be 0.92. Becuase of its simplicity and light weight, a stretched-membrane heliostat could be about 300 less costly than a glass-mirror design.

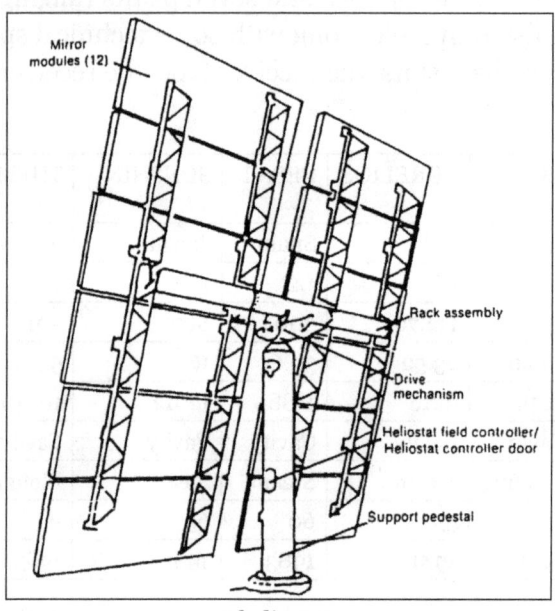

A heliostat.

The receiver is the most complex part of the collection system. The main factor influencing its design is its ability to accept the large and variable heat flux which results from the concentration of the solar radiation by the heliostats. This flux has to be transferred to the receiver fluid. The value of the heat flux can range from 100 to 1000 kW/m² and these results in high temperatures, high thermal gradients and high stresses in the receiver. The value depends on the concentration ratio and varies with the season and the day. It also varies over the surface of the receiver. For these reasons, attention has to be given to the absorber shape, the heat transfer fluid, the arrangement of tubes to carry the fluid and the materials of construction.

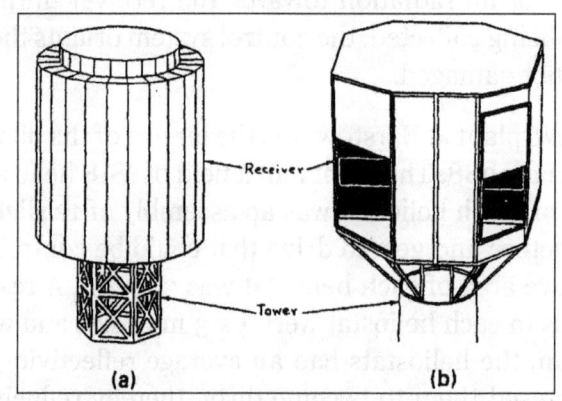

Receivers (a) External type, (b) Cavity type with four apertures.

There are two types of receiver designs: the external type and the cavity type. The external receiver is usually cylindrical in shape. The solar flux is directed onto the outer surface of the cylinder consisting of a number of panels and is absorbed by the receiver fluid flowing through closely spaced tubes fixed on the inner side. On the other hand, in a cavity receiver, the solar flux enters through a small aperture in an insulated enclosure. The cavity contains a suitable tube configuration through which the receiver fluid flows. The geometry of the cavity is such that it maximises the absorption of the entering radiation, minimizes heat losses by convection and radiation to the ambient and at the same time accommodates the heat exchanger that transfers the radiant energy to the receiver fluid. Both types of receivers have their advantages and disadvantages. The external type has a very wide acceptance angle, while the cavity type has a small acceptance angle. On the other hand, the cavity type traps the solar flux more effectively and consequently has a higher efficiency than the external type. The 10 MWe plant at Barstow had an external type of receiver in which water was heated directly and converted to superheated steam. The receiver was a cylinder, 7 m in diameter and 13.5 m in height, made up of 24 vertical panels painted black. Incoloy 800 tubes (0.6 cm I. D., 1.25 cm O. D.) were fixed on the inside. The receiver was located on a tower 80 m high and produced steam at 102 bar and 510 °C. The receiver had an annual efficiency of 0.69, which was rather low.

References

- Basic-electronics-energy-bands, basic-electronics: tutorialspoint.com, Retrieved 09, May 2020
- Solar-thermal-power-generation-technology-research-337864553: researchgate.net, Retrieved 17, January 2020

Chapter 4

Solar Power and Photovoltaics

Simply put, solar power is the use of the sun's energy either directly as thermal energy or its conversion to electricity through the use of photovoltaic cells. The conversion of the sun's energy into electricity is known as photovoltaics. The set-up used to undertake this process of electricity generation is called a photovoltaic system. This chapter has been carefully written to provide the reader with a better understanding of the subject matter.

Solar Power

Solar power is the conversion of sunlight into electricity, either directly using photovoltaics (PV), or indirectly using concentrated solar power. Concentrated solar power systems (Unified Solar) use lenses or mirrors and tracking systems to focus a large area of sunlight into a small beam. Photovoltaics convert light into an electric current using the photovoltaic effect.

The International Energy Agency projected in 2014 that under its "high renewables" scenario, by 2050, solar photovoltaics and concentrated solar power would contribute about 16 and 11 percent, respectively, of the worldwide electricity consumption, and solar would be the world's largest source of electricity. Most solar installations would be in China and India.

Photovoltaics were initially solely used as a source of electricity for small and medium-sized applications, from the calculator powered by a single solar cell to remote homes powered by an off-grid rooftop PV system. As the cost of solar electricity has fallen, the number of grid-connected solar PV systems has grown into the millions and utility-scale solar power stations with hundreds of megawatts are being built. Solar PV is rapidly becoming an inexpensive, low-carbon technology to harness renewable energy from the Sun.

A solar PV array on a rooftop in Hong Kong.

Commercial concentrated solar power plants were first developed in the 1980s. The 392 MW Ivanpah installation is the largest concentrating solar power plant in the world, located in the Mojave Desert of California.

Mainstream Technologies

Many industrialized nations have installed significant solar power capacity into their grids to supplement or provide an alternative to conventional energy sources while an increasing number of less developed nations have turned to solar to reduce dependence on expensive imported fuels. Long distance transmission allows remote renewable energy resources to displace fossil fuel consumption. Solar power plants use one of two technologies:

- Photovoltaic (PV) systems use solar panels, either on rooftops or in ground-mounted solar farms, converting sunlight directly into electric power.

- Concentrated solar power (CSP, also known as "concentrated solar thermal") plants use solar thermal energy to make steam, that is thereafter converted into electricity by a turbine.

Photovoltaics

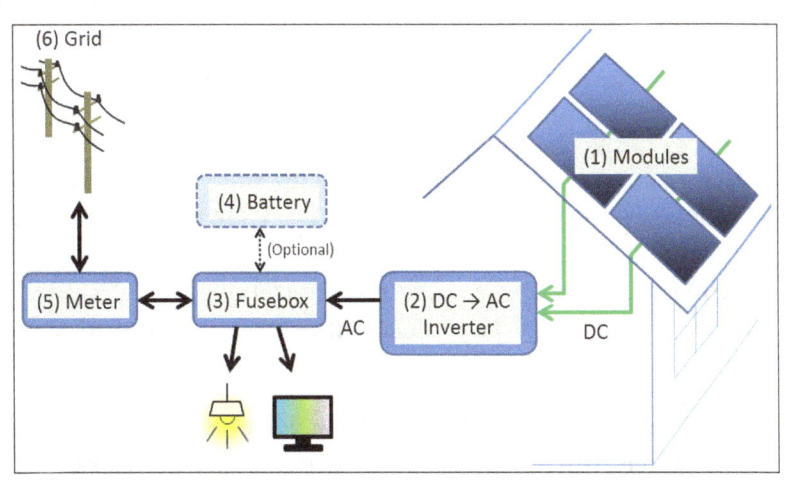

Schematics of a grid-connected residential PV power system.

A solar cell, or photovoltaic cell (PV), is a device that converts light into electric current using the photovoltaic effect. The first solar cell was constructed by Charles Fritts in the 1880s. The German industrialist Ernst Werner von Siemens was among those who recognized the importance of this discovery. In 1931, the German engineer Bruno Lange developed a photo cell using silver selenide in place of copper oxide, although the prototype selenium cells converted less than 1% of incident light into electricity. Following the work of Russell Ohl in the 1940s, researchers Gerald Pearson, Calvin Fuller and Daryl Chapin created the silicon solar cell in 1954. These early solar cells cost 286 USD/watt and reached efficiencies of 4.5–6%.

Conventional PV Systems

The array of a photovoltaic power system, or PV system, produces direct current (DC) power which fluctuates with the sunlight's intensity. For practical use this usually requires conversion to certain desired voltages or alternating current (AC), through the use of inverters. Multiple solar cells are connected inside modules. Modules are wired together to form arrays, then tied to an inverter, which produces power at the desired voltage, and for AC, the desired frequency/phase.

Many residential PV systems are connected to the grid wherever available, especially in developed countries with large markets. In these grid-connected PV systems, use of energy storage is optional. In certain applications such as satellites, lighthouses, or in developing countries, batteries or additional power generators are often added as back-ups. Such stand-alone power systems permit operations at night and at other times of limited sunlight.

Concentrated Solar Power

Concentrated solar power (CSP), also called "concentrated solar thermal", uses lenses or mirrors and tracking systems to focus a large area of sunlight into a small beam. Contrary to photovoltaics – which converts light directly into electricity – CSP uses the heat of the sun's radiation to generate electricity from conventional steam-driven turbines.

A wide range of concentrating technologies exists: among the best known are the parabolic trough, the compact linear Fresnel reflector, the Stirling dish and the solar power tower. Various techniques are used to track the sun and focus light. In all of these systems a working fluid is heated by the concentrated sunlight, and is then used for power generation or energy storage. Thermal storage efficiently allows up to 24-hour electricity generation.

A *parabolic trough* consists of a linear parabolic reflector that concentrates light onto a receiver positioned along the reflector's focal line. The receiver is a tube positioned right above the middle of the parabolic mirror and is filled with a working fluid. The reflector is made to follow the sun during daylight hours by tracking along a single axis. Parabolic trough systems provide the best land-use factor of any solar technology. The SEGS plants in California and Acciona's Nevada Solar One near Boulder City, Nevada are representatives of this technology.

Compact Linear Fresnel Reflectors are CSP-plants which use many thin mirror strips instead of parabolic mirrors to concentrate sunlight onto two tubes with working fluid. This has the advantage that flat mirrors can be used which are much cheaper than parabolic mirrors, and that more reflectors can be placed in the same amount of space, allowing more of the available sunlight to be used. Concentrating linear fresnel reflectors can be used in either large or more compact plants.

The *Stirling solar dish* combines a parabolic concentrating dish with a Stirling engine which normally drives an electric generator. The advantages of Stirling solar over photovoltaic cells are higher efficiency of converting sunlight into electricity and longer lifetime. Parabolic dish systems give the highest efficiency among CSP technologies. The 50 kW Big Dish in Canberra, Australia is an example of this technology.

A *solar power tower* uses an array of tracking reflectors (heliostats) to concentrate light on a central receiver atop a tower. Power towers are more cost effective, offer higher efficiency and better energy storage capability among CSP technologies. The PS10 Solar Power Plant and PS20 solar power plant are examples of this technology.

Hybrid Systems

A hybrid system combines (C)PV and CSP with one another or with other forms of generation such as diesel, wind and biogas. The combined form of generation may enable the system to modulate power output as a function of demand or at least reduce the fluctuating nature of solar power and the consumption of non renewable fuel. Hybrid systems are most often found on islands.

CPV/CSP System:

A novel solar CPV/CSP hybrid system has been proposed, combining concentrator photovoltaics with the non-PV technology of concentrated solar power, or also known as concentrated solar thermal.

ISCC system:

The Hassi R'Mel power station in Algeria, is an example of combining CSP with a gas turbine, where a 25-megawatt CSP-parabolic trough array supplements a much larger 130 MW combined cycle gas turbine plant. Another example is the Yazd power station in Iran.

PVT system:

Hybrid PV/T), also known as photovoltaic thermal hybrid solar collectors convert solar radiation into thermal and electrical energy. Such a system combines a solar (PV) module with a solar thermal collector in an complementary way.

CPVT system:

A concentrated photovoltaic thermal hybrid (CPVT) system is similar to a PVT system. It uses concentrated photovoltaics (CPV) instead of conventional PV technology, and combines it with a solar thermal collector.

PV diesel system:

It combines a photovoltaic system with a diesel generator. Combinations with other renewables are possible and include wind turbines.

PV-thermoelectric system:

Thermoelectric, or "thermovoltaic" devices convert a temperature difference between dissimilar materials into an electric current. Solar cells use only the high frequency part of the radiation, while the low frequency heat energy is wasted. Several patents about the use of thermoelectric devices in tandem with solar cells have been filed. The idea is to increase the efficiency of the combined solar/thermoelectric system to convert the solar radiation into useful electricity.

Development and Deployment

Electricity Generation from Solar		
Year	Energy (TWh)	% of Total
2004	2.6	0.01%
2005	3.7	0.02%
2006	5.0	0.03%
2007	6.8	0.03%
2008	11.4	0.06%
2009	19.3	0.10%
2010	31.4	0.15%
2011	60.6	0.27%
2012	96.7	0.43%
2013	134.5	0.58%
2014	185.9	0.79%
2015	253.0	1.05%

Source: BP-Statistical Review of World Energy, 2016

Early Days

The early development of solar technologies starting in the 1860s was driven by an expectation that coal would soon become scarce. However, development of solar technologies stagnated in the early 20th century in the face of the increasing availability, economy, and utility of coal and petroleum. In 1974 it was estimated that only six private homes in all of North America were entirely heated or cooled by functional solar power systems. The 1973 oil embargo and 1979 energy crisis caused a reorganization of energy policies around the world and brought renewed attention to developing solar technologies. Deployment strategies focused on incentive programs such as the Federal Photovoltaic Utilization Program in the US and the Sunshine Program in Japan. Other efforts included the formation of research facilities in the United States (SERI, now NREL), Japan (NEDO), and Germany (Fraunhofer–ISE). Between 1970 and 1983 installations of photovoltaic systems grew rapidly, but falling oil prices in the early 1980s moderated the growth of photovoltaics from 1984 to 1996.

Mid-1990s to Early 2010s

In the mid-1990s, development of both, residential and commercial rooftop solar as well as utility-scale photovoltaic power stations, began to accelerate again due to supply issues with oil and natural gas, global warming concerns, and the improving economic position of PV relative to other energy technologies. In the early 2000s, the adoption of feed-in tariffs—a policy mechanism, that gives renewables priority on the grid and defines a fixed price for the generated electricity—lead to a high level of investment security and to a soaring number of PV deployments in Europe.

Current Status

For several years, worldwide growth of solar PV was driven by European deployment, but has since shifted to Asia, especially China and Japan, and to a growing number of countries and regions all over the world, including, but not limited to, Australia, Canada, Chile, India, Israel, Mexico, South Africa, South Korea, Thailand, and the United States.

Worldwide growth of photovoltaics has averaged 40% per year since 2000 and total installed capacity reached 139 GW at the end of 2013 with Germany having the most cumulative installations (35.7 GW) and Italy having the highest percentage of electricity generated by solar PV (7.0%).

Concentrated solar power (CSP) also started to grow rapidly, increasing its capacity nearly tenfold from 2004 to 2013, albeit from a lower level and involving fewer countries than solar PV. As of the end of 2013, worldwide cumulative CSP-capacity reached 3,425 MW.

Forecasts

In 2010, the International Energy Agency predicted that global solar PV capacity could reach 3,000 GW or 11% of projected global electricity generation by 2050—enough to generate 4,500 TWh of electricity. Four years later, in 2014, the agency projected that, under its "high renewables" scenario, solar power could supply 27% of global electricity generation by 2050 (16% from PV and 11% from CSP). In 2015, analysts predicted that one million homes in the U.S. will have solar power by the end of 2016.

Photovoltaic Power Stations

The Desert Sunlight Solar Farm is a 550 MW power plant in Riverside County, California, that uses thin-film CdTe-modules made by First Solar. As of November 2014, the 550 megawatt Topaz Solar Farm was the largest photovoltaic power plant in the world. This was surpassed by the 579 MW Solar Star complex. The current largest photovoltaic power station in the world is Longyangxia Dam Solar Park, in Gonghe County, Qinghai, China.

World's largest photovoltaic power stations as of 2015			
Name	Capacity (MW)	Location	Year Completed Info
Longyangxia Dam Solar Park	850	Qinghai, China	2013, 2015
Solar Star I and II	579	USA	2015
Topaz Solar Farm	550	California, USA	2014
Desert Sunlight Solar Farm	550	California, USA	2015
California Valley Solar Ranch	292	California, USA	2013
Agua Caliente Solar Project	290	Arizona, USA	2014
Mount Signal Solar	266	California, USA	2014
Antelope Valley Solar Ranch	266	California, USA	*pending*
Charanka Solar Park	224	Gujarat, India	2012
Mesquite Solar project	207	Arizona, USA	*pending (planned 700 MW)*

Huanghe Hydropower Golmud Solar Park	200	Qinghai, China	2011
Gonghe Industrial Park Phase I	200	China	2013
Imperial Valley Solar Project	200	California, USA	2013
Note: figures rounded. List may change frequently.			

Concentrating Solar Power Stations

Commercial concentrating solar power (CSP) plants, also called "solar thermal power stations", were first developed in the 1980s. The 377 MW Ivanpah Solar Power Facility, located in California's Mojave Desert, is the world's largest solar thermal power plant project. Other large CSP plants include the Solnova Solar Power Station (150 MW), the Andasol solar power station (150 MW), and Extresol Solar Power Station (150 MW), all in Spain. The principal advantage of CSP is the ability to efficiently add thermal storage, allowing the dispatching of electricity over up to a 24-hour period. Since peak electricity demand typically occurs at about 5 pm, many CSP power plants use 3 to 5 hours of thermal storage.

Largest operational solar thermal power stations			
Name	**Capacity (MW)**	**Location**	**Notes**
Ivanpah Solar Power Facility	392	Mojave Desert, California, USA	Operational since February 2014. Located southwest of Las Vegas.
Solar Energy Generating Systems	354	Mojave Desert, California, USA	Commissioned between 1984 and 1991. Collection of 9 units.
Mojave Solar Project	280	Barstow, California, USA	Completed December 2014
Solana Generating Station	280	Gila Bend, Arizona, USA	Completed October 2013. Includes a 6h thermal energy storage
Genesis Solar Energy Project	250	Blythe, California, USA	Completed April 2014
Solaben Solar Power Station	200	Logrosán, Spain	Completed 2012–2013
Noor I	160	Morocco	Completed 2016
Solnova Solar Power Station	150	Seville, Spain	Completed in 2010
Andasol solar power station	150	Granada, Spain	Completed 2011. Includes a 7.5h thermal energy storage.
Extresol Solar Power Station	150	Torre de Miguel Sesmero, Spain	Completed 2010–2012. Extresol 3 includes a 7.5h thermal energy storage

Economics

Cost

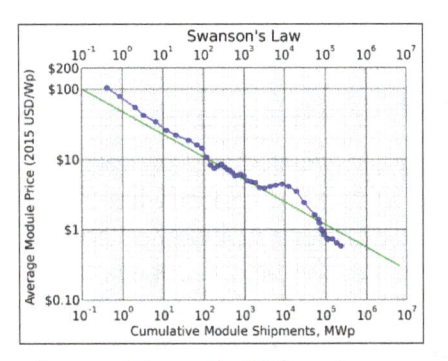

Swanson's law – the PV learning curve.

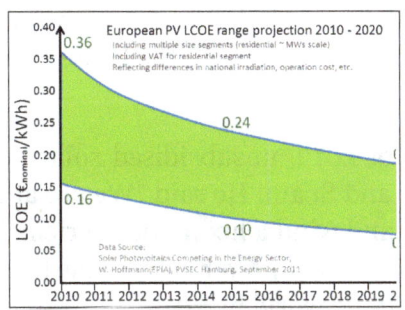

Solar PV – LCOE for Europe until 2020 (in euro-cts. per kWh).

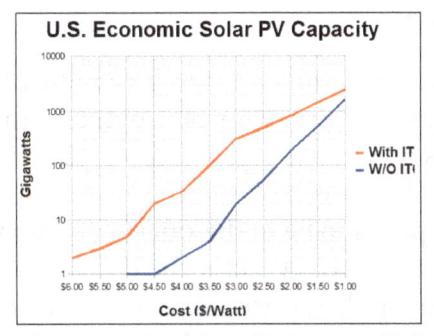

Economic photovoltaic capacity vs installation cost, in the United States.

Adjusting for inflation, it cost $96 per watt for a solar module in the mid-1970s. Process improvements and a very large boost in production have brought that figure down to 68 cents per watt in February 2016, according to data from Bloomberg New Energy Finance. Palo Alto California signed a wholesale purchase agreement in 2016 that secured solar power for 3.7 cents per kilowatt-hour. And in sunny Dubai large-scale solar generated electricity sold in 2016 for just 2.99 cents per kilowatt-hour -- "competitive with any form of fossil-based electricity — and cheaper than most."

Photovoltaic systems use no fuel, and modules typically last 25 to 40 years. Thus, capital costs make up most of the cost of solar power. Operations and maintenance costs for new utility-scale solar plants in the US are estimated to be 9 percent of the cost of photovoltaic electricity, and 17 percent of the cost of solar thermal electricity. Governments have created various financial incentives to encourage the use of solar power, such as feed-in tariff programs. Also, Renewable portfolio standards impose a government mandate that utilities generate or acquire a certain percentage of renewable power regardless of increased energy procurement costs. In most states, RPS goals can be achieved by any combination of solar, wind, biomass, landfill gas, ocean, geothermal, municipal solid waste, hydroelectric, hydrogen, or fuel cell technologies.

Levelized Cost of Electricity

The PV industry is beginning to adopt levelized cost of electricity (LCOE) as the unit of cost. The electrical energy generated is sold in units of kilowatt-hours (kWh). As a rule of thumb, and depending on the local insolation, 1 watt-peak of installed solar PV capacity generates about 1 to 2 kWh of electricity per year. This corresponds to a capacity factor of around 10–20%. The product of the local cost of electricity and the insolation determines the break even point for solar power. The International Conference on Solar Photovoltaic Investments, organized by EPIA, has estimated that PV systems will pay back their investors in 8 to 12 years. As a result, since 2006 it has been economical for investors to install photovoltaics for free in return for a long term power purchase agreement. Fifty percent of commercial systems in the United States were installed in this manner in 2007 and over 90% by 2009.

Shi Zhengrong has said that, as of 2012, unsubsidised solar power is already competitive with fossil fuels in India, Hawaii, Italy and Spain. He said "We are at a tipping point. No longer are renewable power sources like solar and wind a luxury of the rich. They are now starting to compete in the real world without subsidies". "Solar power will be able to compete without subsidies against conventional power sources in half the world by 2015".

Current Installation Prices

In its 2014 edition of the *Technology Roadmap: Solar Photovoltaic Energy* report, the International Energy Agency (IEA) published prices for residential, commercial and utility-scale PV systems for eight major markets as of 2013 *(see table below)*. However, DOE's SunShot Initiative has reported much lower U.S. installation prices. In 2014, prices continued to decline. The SunShot Initiative modeled U.S. system prices to be in the range of $1.80 to $3.29 per watt. Other sources identify similar price ranges of $1.70 to $3.50 for the different market segments in the U.S., and in the highly penetrated German market, prices for residential and small commercial rooftop systems of up to 100 kW declined to $1.36 per watt (€1.24/W) by the end of 2014. In 2015, Deutsche Bank estimated costs for small residential rooftop systems in the U.S. around $2.90 per watt. Costs for utility-scale systems in China and India were estimated as low as $1.00 per watt.

Typical PV system prices in 2013 in selected countries (USD)									
USD/W	Australia	China	France	Germany	Italy	Japan	United Kingdom	United States
Residential	1.8	1.5	4.1	2.4	2.8	4.2	2.8	4.9[1]
Commercial	1.7	1.4	2.7	1.8	1.9	3.6	2.4	4.5[1]
Utility-scale	2.0	1.4	2.2	1.4	1.5	2.9	1.9	3.3[1]

Source: *IEA – Technology Roadmap: Solar Photovoltaic Energy report, September 2014'*
[1] U.S figures are lower in DOE's Photovoltaic System Pricing Trends.

Grid Parity

Grid parity, the point at which the cost of photovoltaic electricity is equal to or cheaper than the price of grid power, is more easily achieved in areas with abundant sun and high costs for electricity such as in California and Japan. In 2008, The levelized cost of electricity for solar PV was $0.25/kWh or less in most of the OECD countries. By late 2011, the fully loaded cost was predicted to fall below $0.15/kWh for most of the OECD and to reach $0.10/kWh in sunnier regions. These cost levels are driving three emerging trends: vertical integration of the supply chain, origination of power purchase agreements (PPAs) by solar power companies, and unexpected risk for traditional power generation companies, grid operators and wind turbine manufacturers.

Grid parity was first reached in Spain in 2013, Hawaii and other islands that otherwise use fossil fuel (diesel fuel) to produce electricity, and most of the US is expected to reach grid parity by 2015.

In 2007, General Electric's Chief Engineer predicted grid parity without subsidies in sunny parts of the United States by around 2015; other companies predicted an earlier date: the cost of solar power will be below grid parity for more than half of residential customers and 10% of commercial customers in the OECD, as long as grid electricity prices do not decrease through 2010.

Self Consumption

In cases of self consumption of the solar energy, the payback time is calculated based on how much electricity is not purchased from the grid. For example, in Germany, with electricity prices of 0.25 Euro/KWh and insolation of 900 KWh/KW, one KWp will save 225 Euro per year, and with an installation cost of 1700 Euro/KWp the system cost will be returned in less than 7 years. However, in many cases, the patterns of generation and consumption do not coincide, and some or all of the energy is fed back into the grid. The electricity is sold, and at other times when energy is taken from the grid, electricity is bought. The relative costs and prices obtained affect the economics.

Energy Pricing and Incentives

The political purpose of incentive policies for PV is to facilitate an initial small-scale deployment to begin to grow the industry, even where the cost of PV is significantly above grid parity, to allow the industry to achieve the economies of scale necessary to reach grid parity. The policies are implemented to promote national energy independence, high tech job creation and reduction of CO_2 emissions. Three incentive mechanisms are often used in combination as investment subsidies: the authorities refund part of the cost of installation of the system, the electricity utility buys PV electricity from the producer under a multiyear contract at a guaranteed rate (), and Solar Renewable Energy Certificates (SRECs).

With investment subsidies, the financial burden falls upon the taxpayer, while with feed-in tariffs the extra cost is distributed across the utilities' customer bases. While the investment subsidy may be simpler to administer, the main argument in favour of feed-in tariffs is the encouragement of quality. Investment subsidies are paid out as a function of the nameplate capacity of the installed system and are independent of its actual power yield over time, thus rewarding the overstatement of power and tolerating poor durability and maintenance. Some electric companies offer rebates to their customers, such as Austin Energy in Texas, which offers $2.50/watt installed up to $15,000.

Net Metering

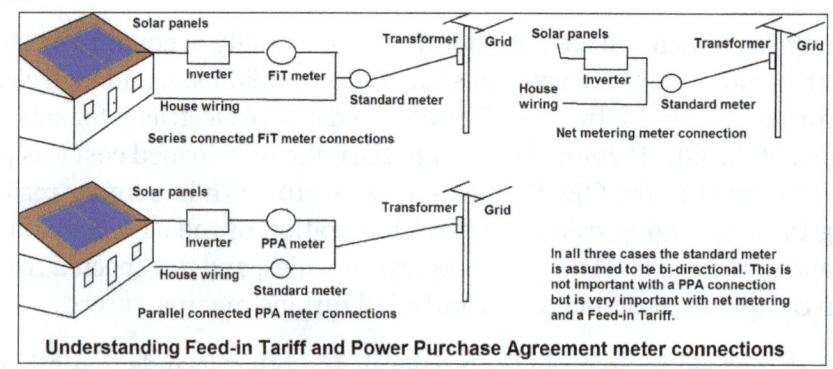

Net metering, unlike a feed-in tariff, requires only one meter, but it must be bi-directional.

In net metering the price of the electricity produced is the same as the price supplied to the consumer, and the consumer is billed on the difference between production and consumption. Net metering can usually be done with no changes to standard electricity meters, which accurately measure power in both directions and automatically report the difference, and because it allows homeowners and businesses to generate electricity at a different time from consumption, effectively using the grid as a giant storage battery. With net metering, deficits are billed each month while surpluses are rolled over to the following month. Best practices call for perpetual roll over of kWh credits. Excess credits upon termination of service are either lost, or paid for at a rate ranging from wholesale to retail rate or above, as can be excess annual credits. In New Jersey, annual excess credits are paid at the wholesale rate, as are left over credits when a customer terminates service.

Feed-In Tariffs (FIT)

With feed-in tariffs, the financial burden falls upon the consumer. They reward the number of kilowatt-hours produced over a long period of time, but because the rate is set by the authorities, it may result in perceived overpayment. The price paid per kilowatt-hour under a feed-in tariff exceeds the price of grid electricity. Net metering refers to the case where the price paid by the utility is the same as the price charged.

The complexity of approvals in California, Spain and Italy has prevented comparable growth to Germany even though the return on investment is better. In some countries, additional incentives are offered for BIPV compared to stand alone PV.

- France + EUR 0.16 /kWh (compared to semi-integrated) or + EUR 0.27/kWh (compared to stand alone)
- Italy + EUR 0.04-0.09 kWh
- Germany + EUR 0.05/kWh (facades only)

Solar Renewable Energy Credits (SRECs)

Alternatively, SRECs allow for a market mechanism to set the price of the solar generated electricity subsity. In this mechanism, a renewable energy production or consumption target is set, and the utility (more technically the Load Serving Entity) is obliged to purchase renewable energy or face a fine (Alternative Compliance Payment or ACP). The producer is credited for an SREC for every 1,000 kWh of electricity produced. If the utility buys this SREC and retires it, they avoid paying the ACP. In principle this system delivers the cheapest renewable energy, since the all solar facilities are eligible and can be installed in the most economic locations. Uncertainties about the future value of SRECs have led to long-term SREC contract markets to give clarity to their prices and allow solar developers to pre-sell and hedge their credits.

Financial incentives for photovoltaics differ across countries, including Australia, China, Germany, Israel, Japan, and the United States and even across states within the US.

The Japanese government through its Ministry of International Trade and Industry ran a successful programme of subsidies from 1994 to 2003. By the end of 2004, Japan led the world in installed PV capacity with over 1.1 GW.

In 2004, the German government introduced the first large-scale feed-in tariff system, under the German Renewable Energy Act, which resulted in explosive growth of PV installations in Germany. At the outset the FIT was over 3x the retail price or 8x the industrial price. The principle behind the German system is a 20-year flat rate contract. The value of new contracts is programmed to decrease each year, in order to encourage the industry to pass on lower costs to the end users. The programme has been more successful than expected with over 1GW installed in 2006, and political pressure is mounting to decrease the tariff to lessen the future burden on consumers.

Subsequently, Spain, Italy, Greece—that enjoyed an early success with domestic solar-thermal installations for hot water needs—and France introduced feed-in tariffs. None have replicated the programmed decrease of FIT in new contracts though, making the German incentive relatively less and less attractive compared to other countries. The French and Greek FIT offer a high premium (EUR 0.55/kWh) for building integrated systems. California, Greece, France and Italy have 30-50% more insolation than Germany making them financially more attractive. The Greek domestic "solar roof" programme (adopted in June 2009 for installations up to 10 kW) has internal rates of return of 10-15% at current commercial installation costs, which, furthermore, is tax free.

In 2006 California approved the 'California Solar Initiative', offering a choice of investment subsidies or FIT for small and medium systems and a FIT for large systems. The small-system FIT of $0.39 per kWh (far less than EU countries) expires in just 5 years, and the alternate "EPBB" residential investment incentive is modest, averaging perhaps 20% of cost. All California incentives are scheduled to decrease in the future depending as a function of the amount of PV capacity installed.

At the end of 2006, the Ontario Power Authority (OPA, Canada) began its Standard Offer Program, a precursor to the Green Energy Act, and the first in North America for distributed renewable projects of less than 10 MW. The feed-in tariff guaranteed a fixed price of $0.42 CDN per kWh over a period of twenty years. Unlike net metering, all the electricity produced was sold to the OPA at the given rate.

Environmental Impacts

Unlike fossil fuel based technologies, solar power does not lead to any harmful emissions during operation, but the production of the panels leads to some amount of pollution.

Part of the Senftenberg Solarpark, a solar photovoltaic power plant located on former open-pit mining areas close to the city of Senftenberg, in Eastern Germany. The 78 MW Phase 1 of the plant was completed within three months.

Greenhouse Gases

The Life-cycle greenhouse-gas emissions of solar power are in the range of 22 to 46 gram (g) per kilowatt-hour (kWh) depending on if solar thermal or solar PV is being analyzed, respectively. With this potentially being decreased to 15 g/kWh in the future. For comparison (of weighted averages), a combined cycle gas-fired power plant emits some 400–599 g/kWh, an oil-fired power plant 893 g/kWh, a coal-fired power plant 915–994 g/kWh or with carbon capture and storage some 200 g/kWh, and a geothermal high-temp. power plant 91–122 g/kWh. The life cycle emission intensity of hydro, wind and nuclear power are lower than solar's as of 2011 as published by the IPCC, and discussed in the article Life-cycle greenhouse-gas emissions of energy sources. Similar to all energy sources were their total life cycle emissions primarily lay in the construction and transportation phase, the switch to low carbon power in the manufacturing and transportation of solar devices would further reduce carbon emissions. BP Solar owns two factories built by Solarex (one in Maryland, the other in Virginia) in which all of the energy used to manufacture solar panels is produced by solar panels. A 1-kilowatt system eliminates the burning of approximately 170 pounds of coal, 300 pounds of carbon dioxide from being released into the atmosphere, and saves up to 105 gallons of water consumption monthly.

The US National Renewable Energy Laboratory (NREL), in harmonizing the disparate estimates of life-cycle GHG emissions for solar PV, found that the most critical parameter was the solar insolation of the site: GHG emissions factors for PV solar are inversely proportional to insolation. For a site with insolation of 1700 kWh/m2/year, typical of southern Europe, NREL researchers estimated GHG emissions of 45 gCO_2e/kWh. Using the same assumptions, at Phoenix, USA, with insolation of 2400 kWh/m2/year, the GHG emissions factor would be reduced to 32 g of CO_2e/kWh.

The New Zealand Parliamentary Commissioner for the Environment found that the solar PV would have little impact on the country's greenhouse gas emissions. The country already generates 80 percent of its electricity from renewable resources (primarily hydroelectricity and geothermal) and national electricity usage peaks on winter evenings whereas solar generation peaks on summer afternoons, meaning a large uptake of solar PV would end up displacing other renewable generators before fossil-fueled power plants.

Energy Payback

The energy payback time (EPBT) of a power generating system is the time required to generate as much energy as is consumed during production and lifetime operation of the system. Due to improving production technologies the payback time has been decreasing constantly since the introduction of PV systems in the energy market. In 2000 the energy payback time of PV systems was estimated as 8 to 11 years and in 2006 this was estimated to be 1.5 to 3.5 years for crystalline silicon silicon PV systems and 1–1.5 years for thin film technologies (S. Europe). These figures fell to 0.75–3.5 years in 2013, with an average of about 2 years for crystalline silicon PV and CIS systems.

Another economic measure, closely related to the energy payback time, is the energy returned on energy invested (EROEI) or energy return on investment (EROI), which is the ratio of electricity generated divided by the energy required to build *and maintain* the equipment. (This is not the same as the economic return on investment (ROI), which varies according to local energy prices, subsidies available and metering techniques.) With expected lifetimes of 30 years, the EROEI of PV systems are in the range of 10 to 30, thus generating enough energy over their lifetimes to reproduce themselves many times (6-31 reproductions) depending on what type of material, balance of system (BOS), and the geographic location of the system.

Other Issues

One issue that has often raised concerns is the use of cadmium (Cd), a toxic heavy metal that has the tendency to accumulate in ecological food chains. It is used as semiconductor component in CdTe solar cells and as buffer layer for certain CIGS cells in the form of CdS. The amount of cadmium used in thin-film PV modules is relatively small (5–10 g/m^2) and with proper recycling and emission control techniques in place the cadmium emissions from module production can be almost zero. Current PV technologies lead to cadmium emissions of 0.3–0.9 microgram/kWh over the whole life-cycle. Most of these emissions actually arise through the use of coal power for the manufacturing of the modules, and coal and lignite combustion leads to much higher emissions of cadmium. Life-cycle cadmium emissions from coal is 3.1 microgram/kWh, lignite 6.2, and natural gas 0.2 microgram/kWh.

In a life-cycle analysis it has been noted, that if electricity produced by photovoltaic panels were used to manufacture the modules instead of electricity from burning coal, cadmium emissions from coal power usage in the manufacturing process could be entirely eliminated.

In the case of crystalline silicon modules, the solder material, that joins together the copper strings of the cells, contains about 36 percent of lead (Pb). Moreover, the paste used for screen printing front and back contacts contains traces of Pb and sometimes Cd as well. It is estimated that about 1,000 metric tonnes of Pb have been used for 100 gigawatts of c-Si solar modules. However, there is no fundamental need for lead in the solder alloy.

Some media sources have reported that concentrated solar power plants have injured or killed large numbers of birds due to intense heat from the concentrated sunrays. This adverse effect does not apply to PV solar power plants, and some of the claims may have been overstated or exaggerated.

A 2014-published life-cycle analysis of land use for various sources of electricity concluded that the large-scale implementation of solar and wind potentially reduces pollution-related environmental impacts. The study found that the land-use footprint, given in square meter-years per megawatt-hour (m²a/MWh), was lowest for wind, natural gas and rooftop PV, with 0.26, 0.49 and 0.59, respectively, and followed by utility-scale solar PV with 7.9. For CSP, the footprint was 9 and 14, using parabolic troughs and solar towers, respectively. The largest footprint had coal-fired power plants with 18 m²a/MWh.

Emerging Technologies

Concentrator Photovoltaics

CPV modules on dual axis solar trackers in Golmud, China.

Concentrator photovoltaics (CPV) systems employ sunlight concentrated onto photovoltaic surfaces for the purpose of electrical power production. Contrary to conventional photovoltaic systems, it uses lenses and curved mirrors to focus sunlight onto small, but highly efficient, multi-junction solar cells. Solar concentrators of all varieties may be used, and these are often mounted on a solar tracker in order to keep the focal point upon the cell as the sun moves across the sky. Luminescent solar concentrators (when combined with a PV-solar cell) can also be regarded as a CPV system. Concentrated photovoltaics are useful as they can improve efficiency of PV-solar panels drastically.

In addition, most solar panels on spacecraft are also made of high efficient multi-junction photovoltaic cells to derive electricity from sunlight when operating in the inner Solar System.

Floatovoltaics

Floatovoltaics are an emerging form of PV systems that float on the surface of irrigation canals, water reservoirs, quarry lakes, and tailing ponds. Several systems exist in France, India, Japan, Korea, the United Kingdom and the United States. These systems reduce the need of valuable land area, save drinking water that would otherwise be lost through evaporation, and show a higher efficiency of solar energy conversion, as the panels are kept at a cooler temperature than they would be on land.

Grid Integration

Since solar energy is not available at night, storing its energy is an important issue in order to have continuous energy availability. Both wind power and solar power are variable renewable energy, meaning that all available output must be taken when it is available, and either stored for *when it can be used later*, or transported over transmission lines to *where it can be used now*. Concentrated solar power plants typically use thermal energy storage to store the solar energy, such as in high-temperature molten salts. These salts are an effective storage medium because they are low-cost, have a high specific heat capacity, and can deliver heat at temperatures compatible with conventional power systems. This method of energy storage is used, for example, by the Solar Two power station, allowing it to store 1.44 TJ in its 68 m³ storage tank, enough to provide full output for close to 39 hours, with an efficiency of about 99%.

Construction of the Salt Tanks which provide efficient thermal energy storage so that output can be provided after the sun goes down, and output can be scheduled to meet demand requirements. The 280 MW Solana Generating Station is designed to provide six hours of energy storage. This allows the plant to generate about 38 percent of its rated capacity over the course of a year.

Thermal energy storage. The Andasol CSP plant uses tanks of molten salt to store solar energy.

Pumped-storage hydroelectricity (PSH). This facility in Geesthacht, Germany, also includes a solar array.

Rechargeable batteries have been traditionally used to store excess electricity in stand alone PV systems. With grid-connected photovoltaic power system, excess electricity can be sent to the electrical grid. Net metering and feed-in tariff programs give these systems a credit for the electricity they produce. This credit offsets electricity provided from the grid when the system cannot meet demand, effectively using the grid as a storage mechanism. Credits are normally rolled over from month to month and any remaining surplus settled annually. When wind and solar are a small fraction of the grid power, other generation techniques can adjust their output appropriately, but as these forms of variable power grow, this becomes less practical. As prices are rapidly declining, PV systems increasingly use rechargeable batteries to store a surplus to be later used at night. Batteries used for grid-storage also stabilize the electrical grid by leveling out peak loads, and play an important role in a smart grid, as they can charge during periods of low demand and feed their stored energy into the grid when demand is high.

Like plug in cars, it is technically possible to have "plug and play" PV. A recent review article found that careful system design would enable such systems to meet all technical and safety requirements.

Common battery technologies used in today's PV systems include, the valve regulated lead-acid battery– a modified version of the conventional lead–acid battery, nickel–cadmium and lithium-ion batteries. Lead-acid batteries are currently the predominant technology used in small-scale, residential PV systems, due to their high reliability, low self discharge and investment and maintenance costs, despite shorter lifetime and lower energy density. However, lithium-ion batteries have the potential to replace lead-acid batteries in the near future, as they are being intensively developed and lower prices are expected due to economies of scale provided by large production facilities such as the Gigafactory 1. In addition, the Li-ion batteries of plug-in electric cars may serve as a future storage devices in a vehicle-to-grid system. Since most vehicles are parked an average of 95 percent of the time, their batteries could be used to let electricity flow from the car to the power lines and back. Other rechargeable batteries used for distributed PV systems include, sodium–sulfur and vanadium redox batteries, two prominent types of a molten salt and a flow battery, respectively.

Conventional hydroelectricity works very well in conjunction with variable electricity sources such as solar and wind, the water can be held back and allowed to flow as required with virtually no energy loss. Where a suitable river is not available, pumped-storage hydroelectricity stores energy in the form of water pumped when surplus electricity is available, from a lower elevation reservoir to a higher elevation one. The energy is recovered when demand is high by releasing the water: the pump becomes a turbine, and the motor a hydroelectric power generator. However, this loses some of the energy to pumpage losses.

The combination of wind and solar PV has the advantage that the two sources complement each other because the peak operating times for each system occur at different times of the day and year. The power generation of such solar hybrid power systems is therefore more constant and fluctuates less than each of the two component subsystems. Solar power is seasonal, particularly in northern/southern climates, away from the equator, suggesting a need for long term seasonal storage in a medium such as hydrogen. The storage requirements vary and in some cases can be met with biomass. The Institute for Solar Energy Supply Technology of the University of Kassel pilot-tested a combined power plant linking solar, wind, biogas and hydrostorage to provide load-following power around the clock, entirely from renewable sources.

Research is also undertaken in this field of artificial photosynthesis. It involves the use of nanotechnology to store solar electromagnetic energy in chemical bonds, by splitting water to produce hydrogen fuel or then combining with carbon dioxide to make biopolymers such as methanol. Many large national and regional research projects on artificial photosynthesis are now trying to develop techniques integrating improved light capture, quantum coherence methods of electron transfer and cheap catalytic materials that operate under a variety of atmospheric conditions. Senior researchers in the field have made the public policy case for a Global Project on Artificial Photosynthesis to address critical energy security and environmental sustainability issues.

Geographic Solar Insolation

Different parts of the world experience different amounts of sunshine, depending on latitude and weather. Locations nearer the equator receive many more hours of sunshine than those further north or south, thus photovoltaic panels can be more economically desirable in some places more than others.

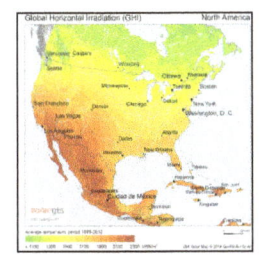

North America.

South America.

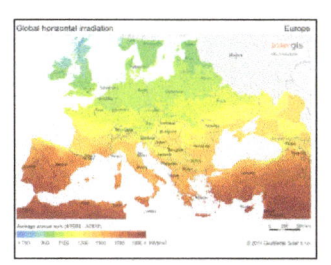

Europe.

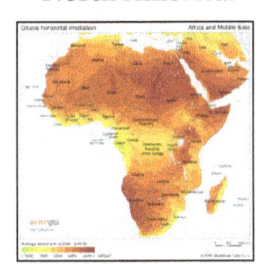

Africa and Middle East.

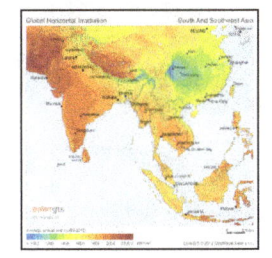

South and South-East Asia.

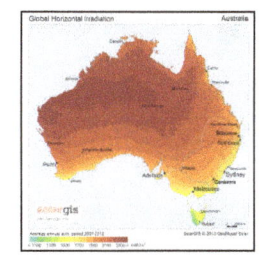

Australia.

Photovoltaics

The Solar Settlement, a sustainable housing community project in Freiburg, Germany.

Photovoltaics (PV) covers the conversion of light into electricity using semiconducting materials that exhibit the photovoltaic effect, a phenomenon studied in physics, photochemistry, and electrochemistry.

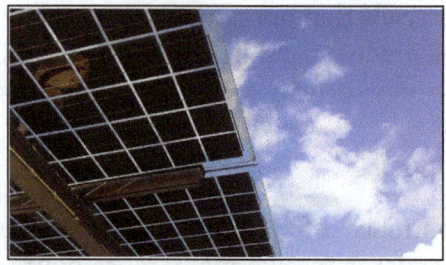

Double glass photovoltaic solar modules, installed in a support structure.

Photovoltaic SUDI shade is an autonomous and mobile station in France that provides energy for electric vehicles using solar energy.

Solar panels on the International Space Station.

A typical photovoltaic system employs solar panels, each comprising a number of solar cells, which generate electrical power. The first step is the photoelectric effect followed by an electrochemical process where crystallized atoms, ionized in a series, generate an electric current. PV Installations may be ground-mounted, rooftop mounted or wall mounted.

Solar PV generates no pollution. The direct conversion of sunlight to electricity occurs without any moving parts. Photovoltaic systems have been used for fifty years in specialized applications, standalone and grid-connected PV systems have been in use for more than twenty years. They were first mass-produced in 2000, when German environmentalists and the Eurosolar organization got government funding for a ten thousand roof program.

On the other hand, grid-connected PV systems have the major disadvantage that the power output is dependent on direct sunlight, so by definition, solar power systems only produce power for half of a day and less if tracking is not used. Power output is also adversely affected by weather conditions, especially cloud cover. This means that, in the national grid for example, this power has to be made up by other power sources: hydrocarbon, nuclear, hydroelectric or wind energy. To some, solar installations also have a negative aesthetic impact on an area.

Advances in technology and increased manufacturing scale have reduced the cost, increased the reliability, and increased the efficiency of photovoltaic instalations and the levelised cost of electricity from PV is competitive, on a kilowatt/ hour basis, with conventional electricity sources in an expanding list of geographic regions. Solar PV regularly costs USD 0.05-0.10 per kilowatt-hour (kWh) in Europe, China, India, South Africa and the United States. In 2015, record low prices were set in the United Arab Emirates (5.84 cents/kWh), Peru (4.8 cents/kWh) and Mexico (4.8 cents/kWh). In May 2016, a solar PV auction in Dubai attracted a bid of 3 cents/kWh.

Net metering and financial incentives, such as preferential feed-in tariffs for solar-generated electricity, have supported solar PV installations in many countries. More than 100 countries now use solar PV. After hydro and wind power, PV is the third renewable energy source in terms of globally capacity. In 2014, worldwide installed PV capacity increased to 177 gigawatts (GW), which is two percent of global electricity demand. China, followed by Japan and the United States, is the fastest growing market, while Germany remains the world's largest producer (both in per capita and absolute terms), with solar PV providing seven percent of annual domestic electricity consumption.

With current technology, photovoltaics recoups the energy needed to manufacture them in 1.5 years in Southern Europe and 2.5 years in Northern Europe.

Etymology

The term "photovoltaic" meaning "light", and from "volt", the unit of electro-motive force, the volt, which in turn comes from the last name of the Italian physi-cist Alessandro Volta, inventor of the battery (electrochemical cell). The term "photo-voltaic" has been in use in English since 1849.

Solar Cells

Solar cells generate electricity directly from sunlight.

Photovoltaics are best known as a method for generating electric power by using solar cells to convert energy from the sun into a flow of electrons. The photovoltaic effect refers to photons of light exciting electrons into a higher state of energy, allowing them to act as charge carriers for an electric current. The photovoltaic effect was first observed by Alexandre-Edmond Becquerel in 1839.

The term photovoltaic denotes the unbiased operating mode of a photodiode in which current through the device is entirely due to the transduced light energy. Virtually all photovoltaic devices are some type of photodiode.

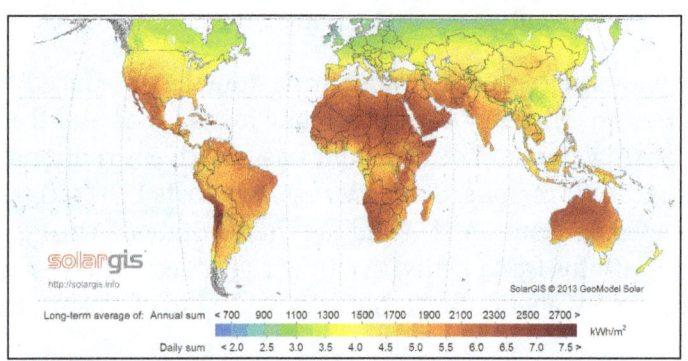

Average insolation. Note that this is for a horizontal surface. Solar panels are normally propped up at an angle and receive more energy per unit area.

Solar cells produce direct current electricity from sun light which can be used to power equipment or to recharge a battery. The first practical application of photovoltaics was to power orbiting satellites and other spacecraft, but today the majority of photovoltaic modules are used for grid connected power generation. In this case an inverter is required to convert the DC to AC. There is a smaller market for off-grid power for remote dwellings, boats, recreational vehicles, electric cars, roadside emergency telephones, remote sensing, and cathodic protection of pipelines.

Photovoltaic power generation employs solar panels composed of a number of solar cells containing a photovoltaic material. Materials presently used for photovoltaics include monocrystalline silicon, polycrystalline silicon, amorphous silicon, cadmium telluride, and copper indium gallium selenide/sulfide. Copper solar cables connect modules (module cable), arrays (array cable), and sub-fields. Because of the growing demand for renewable energy sources, the manufacturing of solar cells and photovoltaic arrays has advanced considerably in recent years.

Solar photovoltaics power generation has long been seen as a clean energy technology which draws upon the planet's most plentiful and widely distributed renewable energy source – the sun. The technology is "inherently elegant" in that the direct conversion of sunlight to electricity occurs without any moving parts or environmental emissions during operation. It is well proven, as photovoltaic systems have now been used for fifty years in specialised applications, and grid-connected systems have been in use for over twenty years.

Cells require protection from the environment and are usually packaged tightly behind a glass sheet. When more power is required than a single cell can deliver, cells are electrically connected together to form photovoltaic modules, or solar panels. A single module is enough to power an emergency telephone, but for a house or a power plant the modules must be arranged in multiples as arrays.

Photovoltaic power capacity is measured as maximum power output under standardized test conditions (STC) in "W_p" (watts peak). The actual power output at a particular point in time may be less than or greater than this standardized, or "rated," value, depending on geographical location, time of day, weather conditions, and other factors. Solar photovoltaic array capacity factors are typically under 25%, which is lower than many other industrial sources of electricity.

Current Developments

For best performance, terrestrial PV systems aim to maximize the time they face the sun. Solar trackers achieve this by moving PV panels to follow the sun. The increase can be by as much as 20% in winter and by as much as 50% in summer. Static mounted systems can be optimized by analysis of the sun path. Panels are often set to latitude tilt, an angle equal to the latitude, but performance can be improved by adjusting the angle for summer or winter. Generally, as with other semiconductor devices, temperatures above room temperature reduce the performance of photovoltaics.

A number of solar panels may also be mounted vertically above each other in a tower, if the zenith distance of the Sun is greater than zero, and the tower can be turned horizontally as a whole and each panels additionally around a horizontal axis. In such a tower the panels can follow the Sun exactly. Such a device may be described as a ladder mounted on a turnable disk. Each step of that ladder is the middle axis of a rectangular solar panel. In case the zenith distance of the Sun reaches zero, the "ladder" may be rotated to the north or the south to avoid a solar panel producing a shadow on a lower solar panel. Instead of an exactly vertical tower one can choose a tower with an axis directed to the polar star, meaning that it is parallel to the rotation axis of the Earth. In this case the angle between the axis and the Sun is always larger than 66 degrees. During a day it is only necessary to turn the panels around this axis to follow the Sun. Installations may be ground-mounted (and sometimes integrated with farming and grazing) or built into the roof or walls of a building (building-integrated photovoltaics).

Another recent development involves the makeup of solar cells. Perovskite is a very inexpensive material which is being used to replace the expensive crystalline silicon which is still part of a standard PV cell build to this day. Michael Graetzel, Director of the Laboratory of Photonics and Interfaces at EPFL says, "Today, efficiency has peaked at 18 percent, but it's expected to get even higher in the future." This is a significant claim, as 20% efficiency is typical among solar panels which use more expensive materials.

Efficiency

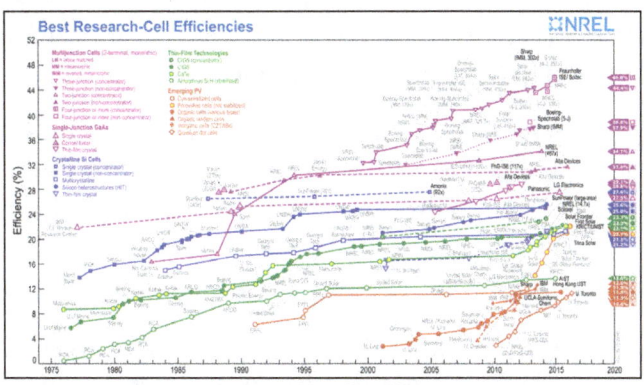

Best Research-Cell Efficiencies.

Electrical efficiency (also called conversion efficiency) is a contributing factor in the selection of a photovoltaic system. However, the most efficient solar panels are typically the most expensive, and may not be commercially available. Therefore, selection is also driven by cost efficiency and other factors.

The electrical efficiency of a PV cell is a physical property which represents how much electrical power a cell can produce for a given insolation. The basic expression for maximum efficiency of a photovoltaic cell is given by the ratio of output power to the incident solar power (radiation flux times area).

$$\eta = \frac{P_{max}}{E \cdot A_{cell}}.$$

The efficiency is measured under ideal laboratory conditions and represents the maximum achievable efficiency of the PV material. Actual efficiency is influenced by the output Voltage, current, junction temperature, light intensity and spectrum.

The most efficient type of solar cell to date is a multi-junction concentrator solar cell with an efficiency of 46.0% produced by Fraunhofer ISE in December 2014. The highest efficiencies achieved without concentration include a material by Sharp Corporation at 35.8% using a proprietary triple-junction manufacturing technology in 2009, and Boeing Spectrolab (40.7% also using a triple-layer design). The US company SunPower produces cells that have an efficiency of 21.5%, well above the market average of 12–18%.

There is an ongoing effort to increase the conversion efficiency of PV cells and modules, primarily for competitive advantage. In order to increase the efficiency of solar cells, it is important to choose a semiconductor material with an appropriate band gap that matches the solar spectrum. This will enhance the electrical and optical properties. Improving the method of charge collection is also useful for increasing the efficiency. There are several groups of materials that are being developed. Ultrahigh-efficiency devices ($\eta>30\%$) are made by using GaAs and GaInP2 semiconductors with multijunction tandem cells. High-quality, single-crystal silicon materials are used to achieve high-efficiency, low cost cells ($\eta>20\%$).

Recent developments in Organic photovoltaic cells (OPVs) have made significant advancements in power conversion efficiency from 3% to over 15% since their introduction in the 1980s. To date, the highest reported power conversion efficiency ranges from 6.7% to 8.94% for small molecule, 8.4%–10.6% for polymer OPVs, and 7% to 21% for perovskite OPVs. OPV's are expected to play a major role in the PV market. Recent improvements have increased the efficiency and lowered cost, while remaining environmentally-benign and renewable.

Several companies have begun embedding power optimizers into PV modules called smart modules. These modules perform maximum power point tracking (MPPT) for each module individually, measure performance data for monitoring, and provide additional safety features. Such modules can also compensate for shading effects, wherein a shadow falling across a section of a module causes the electrical output of one or more strings of cells in the module to decrease.

One of the major causes for the decreased performance of cells is overheating. The efficiency of a solar cell declines by about 0.5% for every 1 degree Celsius increase in temperature. This means that a 100 degree increase in surface temperature could decrease the efficiency of a solar cell by about half. Self-cooling solar cells are one solution to this problem. Rather than using energy to cool the surface, pyramid and cone shapes can be formed from silica, and attached to the surface of a solar panel. Doing so allows visible light to reach the solar cells, but reflects infrared rays (which carry heat).

Growth

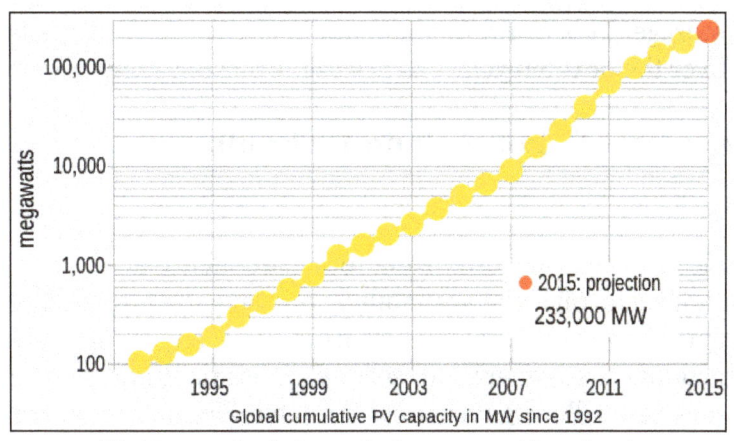

Worldwide growth of photovoltaics on a semi-log plot since 1992.

Solar photovoltaics is growing rapidly and worldwide installed capacity reached at least 177 gigawatts (GW) by the end of 2014. The total power output of the world's PV capacity in a calendar year is now beyond 200 TWh of electricity. This represents 1% of worldwide electricity demand. More than 100 countries use solar PV. China, followed by Japan and the United States is now the fastest growing market, while Germany remains the world's largest producer, contributing more than 7% to its national electricity demands. Photovoltaics is now, after hydro and wind power, the third most important renewable energy source in terms of globally installed capacity.

Several market research and financial companies foresee record-breaking global installation of more than 50 GW in 2015. China is predicted to take the lead from Germany and to become the world's largest producer of PV power by installing another targeted 17.8 GW in 2015. India is expected to install 1.8 GW, doubling its annual installations. By 2018, worldwide photovoltaic capacity is projected to doubled or even triple to 430 GW. Solar Power Europe (formerly known as EPIA) also estimates that photovoltaics will meet 10% to 15% of Europe's energy demand in 2030.

The EPIA/Greenpeace Solar Generation Paradigm Shift Scenario (formerly called Advanced Scenario) from 2010 shows that by the year 2030, 1,845 GW of PV systems could be generating approximately 2,646 TWh/year of electricity around the world. Combined with energy use efficiency improvements, this would represent the electricity needs of more than 9% of the world's population. By 2050, over 20% of all electricity could be provided by photovoltaics.

Michael Liebreich, from Bloomberg New Energy Finance, anticipates a tipping point for solar energy. The costs of power from wind and solar are already below those of conventional electricity generation in some parts of the world, as they have fallen sharply and will continue to do so. He also asserts, that the electrical grid has been greatly expanded worldwide, and is ready to receive and distribute electricity from renewable sources. In addition, worldwide electricity prices came under strong pressure from renewable energy sources, that are, in part, enthusiastically embraced by consumers.

Deutsche Bank sees a "second gold rush" for the photovoltaic industry to come. Grid parity has already been reached in at least 19 markets by January 2014. Photovoltaics will prevail beyond feed-in tariffs, becoming more competitive as deployment increases and prices continue to fall.

In June 2014 Barclays downgraded bonds of U.S. utility companies. Barclays expects more competition by a growing self-consumption due to a combination of decentralized PV-systems and residential electricity storage. This could fundamentally change the utility's business model and transform the system over the next ten years, as prices for these systems are predicted to fall.

Environmental Impacts of Photovoltaic Technologies

Types Of Impacts

While solar photovoltaic (PV) cells are promising for clean energy production, their deployment is hindered by production costs, material availability, and toxicity. Life cycle assessment (LCA) is one method of determining environmental impacts from PV. Many studies have been done on the various types of PV including first generation, second generation, and third generation. Usually these PV LCA studies select a cradle to gate system boundary because often at the time the studies are conducted, it is a new technology not commercially available yet and their required balance of system components and disposal methods are unknown.

A traditional LCA can look at many different impact categories ranging from global warming potential, eco-toxicity, human toxicity, water depletion, and many others. Most LCAs of PV have focused on two categories: carbon dioxide equivalents per kWh and energy pay-back time (EPBT). The EPBT is defined as " the time needed to compensate for the total renewable- and non-renewable- primary energy required during the life cycle of a PV system". A 2015 review of EPBT from first and second generation PV suggested that there was greater variation in embedded energy than in efficiency of the cells implying that it was mainly the embedded energy that needs to reduce to have a greater reduction in EPBT. One difficulty in determining impacts due to PV is to determine if the wastes are released to the air, water, or soil during the manufacturing phase. Research is underway to try to understand emissions and releases during the lifetime of PV systems.

Impacts from First-Generation PV

Crystalline silicon modules are the most extensively studied PV type in terms of LCA since they are the most commonly used. Mono-crystalline silicon photovoltaic systems (mono-si) have an average efficiency of 14.0%. The cells tend to follow a structure of front electrode, anti-reflection film, n-layer, p-layer, and back electrode, with the sun hitting the front electrode. EPBT ranges from 1.7 to 2.7 years. The cradle to gate of CO_2-eq/kWh ranges from 37.3 to 72.2 grams.

Techniques to produce multi-crystalline silicon (multi-si) photovoltaic cells are simpler and cheaper than mono-si, however tend to make less efficient cells, an average of 13.2%. EPBT ranges from 1.5 to 2.6 years. The cradle to gate of CO_2-eq/kWh ranges from 28.5 to 69 grams. Some studies have looked beyond EPBT and GWP to other environmental impacts. In one such study, conventional energy mix in Greece was compared to multi-si PV and found a 95% overall reduction in impacts including carcinogens, eco-toxicity, acidification, eutrophication, and eleven others.

Impacts from Second Generation

Cadmium telluride (CdTe) is one of the fastest-growing thin film based solar cells which are collectively known as second generation devices. This new thin film device also shares similar performance restrictions (Shockley-Queisser efficiency limit) as conventional Si devices but promises to lower the cost of each device by both reducing material and energy consumption during manufacturing. Today the global market share of CdTe is 5.4%, up from 4.7% in 2008. This technology's highest power conversion efficiency is 21%. The cell structure includes glass substrate (around 2 mm), transparent conductor layer, CdS buffer layer (50–150 nm), CdTe absorber and a metal contact layer.

CdTe PV systems require less energy input in their production than other commercial PV systems per unit electricity production. The average CO_2-eq/kWh is around 18 grams (cradle to gate). CdTe has the fastest EPBT of all commercial PV technologies, which varies between 0.3 and 1.2 years.

Copper Indium Gallium Diselenide (CIGS) is a thin film solar cell based on the copper indium diselenide (CIS) family of chalcopyrite semiconductors. CIS and CIGS are often used interchangeably within the CIS/CIGS community. The cell structure includes soda lime glass as the substrate, Mo layer as the back contact, CIS/CIGS as the absorber layer, cadmium sulfide (CdS) or Zn (S,OH)x as the buffer layer, and ZnO:Al as the front contact. CIGS is approximately 1/100th the thickness of conventional silicon solar cell technologies. Materials necessary for assembly are readily available, and are less costly per watt of solar cell. CIGS based solar devices resist performance degradation over time and are highly stable in the field.

Reported global warming potential impacts of CIGS range from 20.5 – 58.8 grams CO_2-eq/kWh of electricity generated for different solar irradiation (1,700 to 2,200 kWh/m²/y) and power conversion efficiency (7.8 – 9.12%). EPBT ranges from 0.2 to 1.4 years, while harmonized value of EPBT was found 1.393 years. Toxicity is an issue within the buffer layer of CIGS modules because it contains cadmium and gallium. CIS modules do not contain any heavy metals.

Impacts from Third Generation

Third-generation PVs are designed to combine the advantages of both the first and second generation devices and they do not have Shockley-Queisser efficiency limit, a theoretical limit for first and second generation PV cells. The thickness of a third generation device is less than 1 μm.

One emerging alternative and promising technology is based on an organic-inorganic hybrid solar cell made of methylammonium lead halide perovskites. Perovskite PV cells have progressed rapidly over the past few years and have become one of the most attractive areas for PV research. The cell structure includes a metal back contact (which can be made of Al, Au or Ag), a hole transfer layer (spiro-MeOTAD, P3HT, PTAA, CuSCN, CuI, or NiO), and absorber layer ($CH_3NH_3PbI_xBr_3$-x, $CH_3NH_3PbI_xCl_3$-x or $CH_3NH_3PbI_3$), an electron transport layer (TiO, ZnO, Al_2O_3 or SnO_2) and a top contact layer (fluorine doped tin oxide or tin doped indium oxide).

There are a limited number of published studies to address the environmental impacts of perovskite solar cells. The major environmental concern is the lead used in the absorber layer. Due to the instability of perovskite cells lead may eventually be exposed to fresh water during the use phase. Two published LCA studies looked at human and ecotoxicity of perovskite solar cells and found they were surprisingly low and may not be an environmental issue. Gong et al. found direct processing energy as 30 MJ/m², while Espinosa didn't report this value (but estimated around 1000 MJ/m²). Global warming potential was found to be in the range of 24–1500 grams CO_2-eq/kWh electricity production. Similarly, reported EPBT of the published paper range from 0.2 to 15 years. The large range of reported values highlight the uncertainties associated with these studies.

Two new promising thin film technologies are copper zinc tin sulfide (Cu$_2$ZnSnS$_4$ or CZTS) and zinc phosphide (Zn$_3$P$_2$). Both of these thin films are currently only produced in the lab but may be commercialized in the future. Their manufacturing processes are expected to be similar to those of current thin film technologies of CIGS and CdTe, respectively. Yet, contrary to CIGS and CdTe, CZTS and Zn$_3$P$_2$ are made from earth abundant, nontoxic materials and have the potential to produce more electricity annually than the current worldwide consumption. While CZTS and Zn$_3$P$_2$ offer good promise for these reasons, the specific environmental implications of their commercial production are not yet known. Global warming potential of CZTS and Zn$_3$P$_2$ were found 38 and 30 grams CO$_2$-eq/kWh while their corresponding EPBT were found 1.85 and 0.78 years, respectively. Overall, CdTe and Zn$_3$P$_2$ have similar environmental impacts but can slightly outperform CIGS and CZTS.

Organic and polymer photovoltaic (OPV) are a relatively new area of research. The tradition OPV cell structure layers consist of a semi-transparent electrode, electron blocking layer, tunnel junction, holes blocking layer, electrode, with the sun hitting the transparent electrode. OPV replaces silver with carbon as an electrode material lowering manufacturing cost and making them more environmentally friendly. OPV are flexible, low weight, and work well with roll-to roll manufacturing for mass production. OPV uses "only abundant elements coupled to an extremely low embodied energy through very low processing temperatures using only ambient processing conditions on simple printing equipment enabling energy pay-back times". Current efficiencies range from 1–6.5%, however theoretical analyses show promise beyond 10% efficiency.

Many different configurations of OPV exist using different materials for each layer. OPV technology rivals existing PV technologies in terms of EPBT even if they currently present a shorter operational lifetime. A 2013 study analyzed 12 different configurations all with 2% efficiency, the EPBT ranged from 0.29–0.52 years for 1 m² of PV. The average CO$_2$-eq/kWh for OPV is 54.922 grams.

Economics

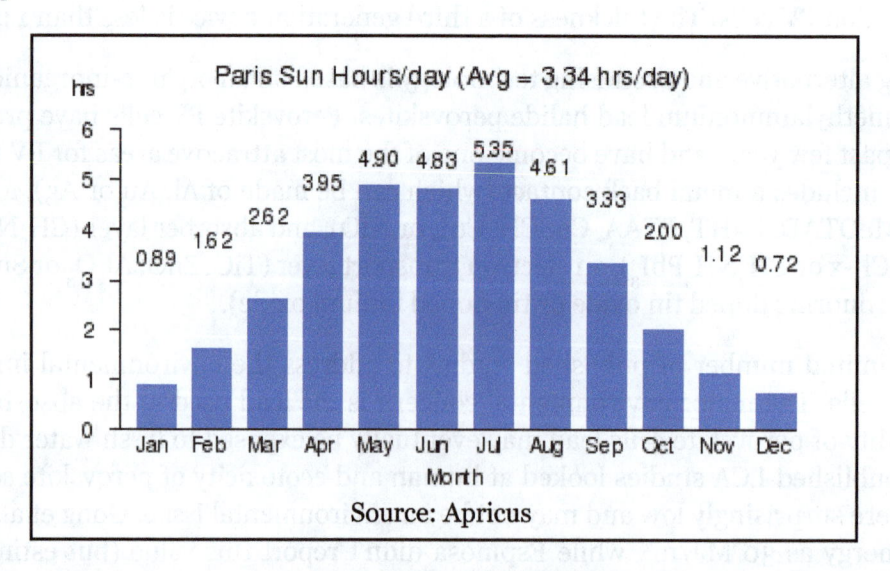

Paris Sun Hours/day (Avg = 3.34 hrs/day)

Month	hrs
Jan	0.89
Feb	1.62
Mar	2.62
Apr	3.95
May	4.90
Jun	4.83
Jul	5.35
Aug	4.61
Sep	3.33
Oct	2.00
Nov	1.12
Dec	0.72

Source: Apricus

There have been major changes in the underlying costs, industry structure and market prices of solar photovoltaics technology, over the years, and gaining a coherent picture of the shifts occurring across the industry value chain globally is a challenge. This is due to: "the rapidity of cost and price changes, the complexity of the PV supply chain, which involves a large number of manufacturing processes, the balance of system (BOS) and installation costs associated with complete PV systems, the choice of different distribution channels, and differences between regional markets within which PV is being deployed". Further complexities result from the many different policy support initiatives that have been put in place to facilitate photovoltaics commercialisation in various countries.

The PV industry has seen dramatic drops in module prices since 2008. In late 2011, factory-gate prices for crystalline-silicon photovoltaic modules dropped below the $1.00/W mark. The $1.00/W installed cost, is often regarded in the PV industry as marking the achievement of grid parity for PV. Technological advancements, manufacturing process improvements, and industry re-structuring, mean that further price reductions are likely in coming years.

Financial incentives for photovoltaics, such as feed-in tariffs, have often been offered to electricity consumers to install and operate solar-electric generating systems. Government has sometimes also offered incentives in order to encourage the PV industry to achieve the economies of scale needed to compete where the cost of PV-generated electricity is above the cost from the existing grid. Such policies are implemented to promote national or territorial energy independence, high tech job creation and reduction of carbon dioxide emissions which cause global warming. Due to economies of scale solar panels get less costly as people use and buy more—as manufacturers increase production to meet demand, the cost and price is expected to drop in the years to come.

Solar cell efficiencies vary from 6% for amorphous silicon-based solar cells to 44.0% with multiple-junction concentrated photovoltaics. Solar cell energy conversion efficiencies for commercially available photovoltaics are around 14–22%. Concentrated photovoltaics (CPV) may reduce cost by concentrating up to 1,000 suns (through magnifying lens) onto a smaller sized photovoltaic cell. However, such concentrated solar power requires sophisticated heat sink designs, otherwise the photovoltaic cell overheats, which reduces its efficiency and life. To further exacerbate the concentrated cooling design, the heat sink must be passive, otherwise the power required for active cooling would reduce the overall efficiency and economy.

Crystalline silicon solar cell prices have fallen from $76.67/Watt in 1977 to an estimated $0.74/Watt in 2013. This is seen as evidence supporting Swanson's law, an observation similar to the famous Moore's Law that states that solar cell prices fall 20% for every doubling of industry capacity.

As of 2011, the price of PV modules has fallen by 60% since the summer of 2008, according to Bloomberg New Energy Finance estimates, putting solar power for the first time on a competitive footing with the retail price of electricity in a number of sunny countries; an alternative and consistent price decline figure of 75% from 2007 to 2012 has also been published, though it is unclear whether these figures are specific to the United States or generally global. The levelised cost of electricity (LCOE) from PV is competitive with conventional electricity sources in an expanding list of geographic regions, particularly when the time of generation is included, as electricity is worth more during the day than at night. There has been fierce competition in the supply chain, and further improvements in the levelised cost of energy for solar lie ahead, posing a growing threat to the dominance of fossil fuel generation sources in the next few years. As time progresses, renewable energy technologies generally get cheaper, while fossil fuels generally get more expensive:

The less solar power costs, the more favorably it compares to conventional power, and the more attractive it becomes to utilities and energy users around the globe. Utility-scale solar power can now be delivered in California at prices well below $100/MWh ($0.10/kWh) less than most other peak generators, even those running on low-cost natural gas. Lower solar module costs also stimulate demand from consumer markets where the cost of solar compares very favorably to retail electric rates.

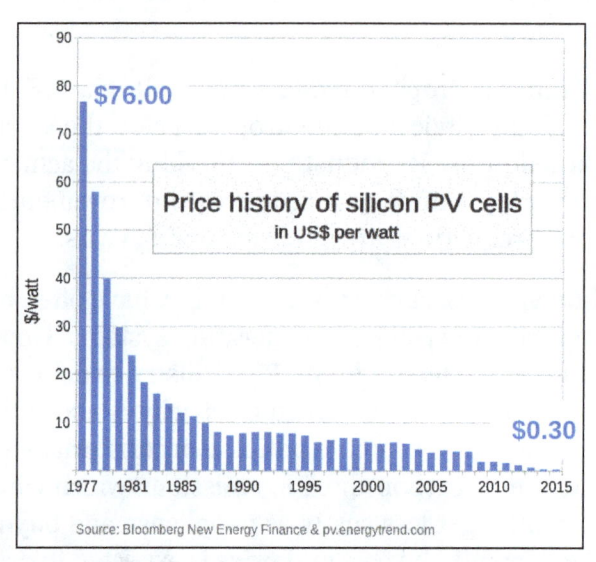

Price per watt history for conventional (c-Si) solar cells since 1977.

As of 2011, the cost of PV has fallen well below that of nuclear power and is set to fall further. The average retail price of solar cells as monitored by the Solarbuzz group fell from $3.50/watt to $2.43/watt over the course of 2011.

For large-scale installations, prices below $1.00/watt were achieved. A module price of 0.60 Euro/watt ($0.78/watt) was published for a large scale 5-year deal in April 2012.

By the end of 2012, the "best in class" module price had dropped to $0.50/watt, and was expected to drop to $0.36/watt by 2017.

In many locations, PV has reached grid parity, which is usually defined as PV production costs at or below retail electricity prices (though often still above the power station prices for coal or gas-fired generation without their distribution and other costs). However, in many countries there is still a need for more access to capital to develop PV projects. To solve this problem securitization has been proposed and used to accelerate development of solar photovoltaic projects. For example, SolarCity offered, the first U.S. asset-backed security in the solar industry in 2013.

Photovoltaic power is also generated during a time of day that is close to peak demand (precedes it) in electricity systems with high use of air conditioning. More generally, it is now evident that, given a carbon price of $50/ton, which would raise the price of coal-fired power by 5c/kWh, solar PV will be cost-competitive in most locations. The declining price of PV has been reflected in rapidly growing installations, totaling about 23 GW in 2011. Although some consolidation is likely in 2012, due to support cuts in the large markets of Germany and Italy, strong growth seems likely to continue for the rest of the decade. Already, by one estimate, total investment in renewables for 2011 exceeded investment in carbon-based electricity generation.

In the case of self consumption payback time is calculated based on how much electricity is not brought from the grid. Additionally, using PV solar power to charge DC batteries, as used in Plug-in Hybrid Electric Vehicles and Electric Vehicles, leads to greater efficiencies. Traditionally, DC generated electricity from solar PV must be converted to AC for buildings, at an average 10% loss during the conversion. An additional efficiency loss occurs in the transition back to DC for battery driven devices and vehicles, and using various interest rates and energy price changes were calculated to find present values that range from $2,057.13 to $8,213.64 (analysis from 2009).

For example, in Germany with electricity prices of 0.25 euro/kWh and Insolation of 900 kWh/kW one kW_p will save 225 euro per year and with installation cost of 1700 euro/kW_p means that the system will pay back in less than 7 years.

Manufacturing

Overall the manufacturing process of creating solar photovoltaics is simple in that it does not require the culmination of many complex or moving parts. Because of the solid state nature of PV systems they often have relatively long lifetimes, anywhere from 10 to 30 years. In order to increase electrical output of a PV system the manufacturer must simply add more photovoltaic components and because of this economies of scale are important for manufacturers as costs decrease with increasing output.

While there are many types of PV systems known to be effective, crystalline silicon PV accounted for around 90% of the worldwide production of PV in 2013. Manufacturing silicon PV systems has several steps. First, polysilicon is processed from mined quartz until it is very pure (semi-conductor grade). This is melted down when small amounts of Boron, a group III element, are added to make a p-type semiconductor rich in electron holes. Typically using a seed crystal, an ingot of this solution is grown from the liquid polycrystalline. The ingot may also be cast in a mold. Wafers of this semiconductor material are cut from the bulk material with wire saws, and then go through surface etching before being cleaned. Next, the wafers are placed into a phosphorus vapor deposition furnace which lays a very thin layer of phosphorus, a group V element, which creates an N-type semiconducting surface. To reduce energy losses an anti-reflective coating is added to the surface, along with electrical contacts. After finishing the cell, cells are connected via electrical circuit according to the specific application and prepared for shipping and installation.

Crystalline silicon photovoltaics are only one type of PV, and while they represent the majority of solar cells produced currently there are many new and promising technologies that have the potential to be scaled up to meet future energy needs.

Another newer technology, thin-film PV, are manufactured by depositing semiconducting layers on substrate in vacuum. The substrate is often glass or stainless-steel, and these semiconducting layers are made of many types of materials including cadmium telluride (CdTe), copper indium diselenide (CIS), copper indium gallium diselenide (CIGS), and amorphous silicon (a-Si). After being deposited onto the substrate the semiconducting layers are separated and connected by electrical circuit by laser-scribing. Thin-film photovoltaics now make up around 20% of the overall production of PV because of the reduced materials requirements and cost to manufacture modules consisting of thin-films as compared to silicon-based wafers.

Other emerging PV technologies include organic, dye-sensitized, quantum-dot, and Perovskite photovoltaics. OPVs fall into the thin-film category of manufacturing, and typically operate around the 12% efficiency range which is lower than the 12–21% typically seen by silicon based PVs. Because organic photovoltaics require very high purity and are relatively reactive they must be encapsulated which vastly increases cost of manufacturing and meaning that they are not feasible for large scale up. Dye-sensitized PVs are similar in efficiency to OPVs but are significantly easier to manufacture. However these dye-sensitized photovoltaics present storage problems because the liquid electrolyte is toxic and can potentially permeate the plastics used in the cell. Quantum dot solar cells are quantum dot sensitized DSSCs and are solution processed meaning they are potentially scalable, but currently they have not reached greater than 10% efficiency. Perovskite solar cells are a very efficient solar energy converter and have excellent optoelectric properties for photovoltaic purposes, but they are expensive and difficult to manufacture.

Applications

Photovoltaic Systems

A photovoltaic system, or solar PV system is a power system designed to supply usable solar power by means of photovoltaics. It consists of an arrangement of several components, including solar panels to absorb and directly convert sunlight into electricity, a solar inverter to change the electric current from DC to AC, as well as mounting, cabling and other electrical accessories. PV systems range from small, roof-top mounted or building-integrated systems with capacities from a few to several tens of kilowatts, to large utility-scale power stations of hundreds of megawatts. Nowadays, most PV systems are grid-connected, while stand-alone systems only account for a small portion of the market.

- Rooftop and building integrated systems:

Rooftop PV on half-timbered house.

Photovoltaic arrays are often associated with buildings: either integrated into them, mounted on them or mounted nearby on the ground. Rooftop PV systems are most often retrofitted into existing buildings, usually mounted on top of the existing roof structure or on the existing walls. Alternatively, an array can be located separately from the building but connected by cable to supply power for the building. Building-integrated photovoltaics (BIPV) are increasingly incorporated into the roof or walls of new domestic and industrial buildings as a principal or ancillary source of electrical power. Roof tiles with integrated PV cells are sometimes used as well. Provided there is an open gap in which air can circulate, rooftop mounted solar panels can provide a passive cooling effect on buildings during the day and also keep accumulated heat in at night. Typically, residential rooftop systems have small capacities of around 5–10 kW, while commercial rooftop systems often amount to several hundreds of kilowatts. Although rooftop systems are much smaller than ground-mounted utility-scale power plants, they account for most of the worldwide installed capacity.

- Concentrator photovoltaics:

Concentrator photovoltaics (CPV) is a photovoltaic technology that contrary to conventional flat-plate PV systems uses lenses and curved mirrors to focus sunlight onto small, but highly efficient, multi-junction (MJ) solar cells. In addition, CPV systems often use solar trackers and sometimes a cooling system to further increase their efficiency. Ongoing research and development is rapidly improving their competitiveness in the utility-scale segment and in areas of high solar insolation.

- Photovoltaic thermal hybrid solar collector:

Photovoltaic thermal hybrid solar collector (PVT) are systems that convert solar radiation into thermal and electrical energy. These systems combine a solar PV cell, which converts sunlight into electricity, with a solar thermal collector, which captures the remaining energy and removes waste heat from the PV module. The capture of both electricity and heat allow these devices to have higher exergy and thus be more overall energy efficient than solar PV or solar thermal alone.

- Power stations:

Many utility-scale solar farms have been constructed all over the world. As of 2015, the 579-megawatt (MW_{AC}) Solar Star is the world's largest photovoltaic power station, followed by the Desert Sunlight Solar Farm and the Topaz Solar Farm, both with a capacity of 550 MW_{AC}, constructed by US-company First Solar, using CdTe modules, a thin-film PV technology. All three power stations are located in the Californian desert. Many solar farms around the world are integrated with agriculture and some use innovative solar tracking systems that follow the sun's daily path across the sky to generate more electricity than conventional fixed-mounted systems. There are no fuel costs or emissions during operation of the power stations.

- Rural electrification:

Developing countries where many villages are often more than five kilometers away from grid power are increasingly using photovoltaics. In remote locations in India a rural lighting program has been providing solar powered LED lighting to replace kerosene lamps. The solar powered lamps were sold at about the cost of a few months' supply of kerosene. Cuba is working to provide solar power for areas that are off grid. More complex applications of off-grid solar energy use include 3D printers. RepRap 3D printers have been solar powered with photovoltaic technology, which enables distributed manufacturing for sustainable development. These are areas where the social costs and benefits offer an excellent case for going solar, though the lack of profitability has relegated such endeavors to humanitarian efforts. However, in 1995 solar rural electrification projects had been found to be difficult to sustain due to unfavorable economics, lack of technical support, and a legacy of ulterior motives of north-to-south technology transfer.

- Standalone systems:

Standalone PV system at an ecotourism resort (British Columbia, Canada).

Until a decade or so ago, PV was used frequently to power calculators and novelty devices. Improvements in integrated circuits and low power liquid crystal displays make it possible to power such devices for several years between battery changes, making PV use less common. In contrast, solar powered remote fixed devices have seen increasing use recently in locations where significant connection cost makes grid power prohibitively expensive. Such applications include solar lamps, water pumps, parking meters, emergency telephones, trash compactors, temporary traffic signs, charging stations, and remote guard posts and signals.

- Floatovoltaics:

In May 2008, the Far Niente Winery in Oakville, CA pioneered the world's first "floatovoltaic" system by installing 994 photovoltaic solar panels onto 130 pontoons and floating them on the winery's irrigation pond. The floating system generates about 477 kW of peak output and when combined with an array of cells located adjacent to the pond is able to fully offset the winery's electricity consumption. The primary benefit of a floatovoltaic system is that it avoids the need to sacrifice valuable land area that could be used for another purpose. In the case of the Far Niente Winery, the floating system saved three-quarters of an acre that would have been required for a land-based system. That land area can instead be used for agriculture. Another benefit of a floatovoltaic system is that the panels are kept at a lower temperature than they would be on land, leading to a higher efficiency of solar energy conversion. The floating panels also reduce the amount of water lost through evaporation and inhibit the growth of algae.

- In transport:

Solar Impulse 2, a solar aircraft.

PV has traditionally been used for electric power in space. PV is rarely used to provide motive power in transport applications, but is being used increasingly to provide auxiliary power in boats and cars. Some automobiles are fitted with solar-powered air conditioning to limit interior temperatures on hot days. A self-contained solar vehicle would have limited power and utility, but a solar-charged electric vehicle allows use of solar power for transportation. Solar-powered cars, boats and airplanes have been demonstrated, with the most practical and likely of these being solar cars. The Swiss solar aircraft, Solar Impulse 2, achieved the longest non-stop solo flight in history and plan to make the first solar-powered aerial circumnavigation of the globe in 2015.

- Telecommunication and signaling:

Solar PV power is ideally suited for telecommunication applications such as local telephone exchange, radio and TV broadcasting, microwave and other forms of electronic communication links. This is because, in most telecommunication application, storage batteries are already in use and the electrical system is basically DC. In hilly and mountainous terrain, radio and TV signals may not reach as they get blocked or reflected back due to undulating terrain. At these locations, low power transmitters (LPT) are installed to receive and retransmit the signal for local population.

- Spacecraft applications:

Part of Juno's solar array.

Solar panels on spacecraft are usually the sole source of power to run the sensors, active heating and cooling, and communications. A battery stores this energy for use when the solar panels are in shadow. In some, the power is also used for spacecraft propulsion—electric propulsion. Spacecraft were one of the earliest applications of photovoltaics, starting with the silicon solar cells used on the Vanguard 1 satellite, launched by the US in 1958. Since then, solar power has been used on missions ranging from the MESSENGER probe to Mercury, to as far out in the solar system as the Juno probe to Jupiter. The largest solar power system flown in space is the electrical system of the International Space Station. To increase the power generated per kilogram, typical spacecraft solar panels use high-cost, high-efficiency, and close-packed rectangular multi-junction solar cells made of gallium arsenide (GaAs) and other semiconductor materials.

- Specialty Power Systems:

Photovoltaics may also be incorporated as energy conversion devices for objects at elevated temperatures and with preferable radiative emissivities such as heterogeneous combustors.

Advantages

The 122 PW of sunlight reaching the Earth's surface is plentiful—almost 10,000 times more than the 13 TW equivalent of average power consumed in 2005 by humans. This abundance leads to the suggestion that it will not be long before solar energy will become the world's primary energy source. Additionally, solar electric generation has the highest power density (global mean of 170 W/m^2) among renewable energies.

Solar power is pollution-free during use. Production end-wastes and emissions are manageable using existing pollution controls. End-of-use recycling technologies are under development and policies are being produced that encourage recycling from producers.

PV installations can operate for 100 years or even more with little maintenance or intervention after their initial set-up, so after the initial capital cost of building any solar power plant, operating costs are extremely low compared to existing power technologies.

Grid-connected solar electricity can be used locally thus reducing transmission/distribution losses (transmission losses in the US were approximately 7.2% in 1995).

Compared to fossil and nuclear energy sources, very little research money has been invested in the development of solar cells, so there is considerable room for improvement. Nevertheless, experimental high efficiency solar cells already have efficiencies of over 40% in case of concentrating photovoltaic cells and efficiencies are rapidly rising while mass-production costs are rapidly falling.

In some states of the United States, much of the investment in a home-mounted system may be lost if the home-owner moves and the buyer puts less value on the system than the seller. The city of Berkeley developed an innovative financing method to remove this limitation, by adding a tax assessment that is transferred with the home to pay for the solar panels. Now known as PACE, Property Assessed Clean Energy, 30 U.S. states have duplicated this solution.

There is evidence, at least in California, that the presence of a home-mounted solar system can actually increase the value of a home. According to a paper published in April 2011 by the Ernest Orlando Lawrence Berkeley National Laboratory titled An Analysis of the Effects of Residential Photovoltaic Energy Systems on Home Sales Prices in California.

The research finds strong evidence that homes with PV systems in California have sold for a premium over comparable homes without PV systems. More specifically, estimates for average PV premiums range from approximately $3.9 to $6.4 per installed watt (DC) among a large number of different model specifications, with most models coalescing near $5.5/watt. That value corresponds to a premium of approximately $17,000 for a relatively new 3,100 watt PV system (the average size of PV systems in the study).

Photovoltaic System

A photovoltaic system, also solar PV power system, or PV system, is a power system designed to supply usable solar power by means of photovoltaics. It consists of an arrangement of several components, including solar panels to absorb and convert sunlight into electricity, a solar inverter to change the electric current from DC to AC, as well as mounting, cabling and other electrical accessories to set up a working system. It may also use a solar tracking system to improve the system's overall performance and include an integrated battery solution, as prices for storage devices are expected to decline. Strictly speaking, a solar array only encompasses the ensemble of solar panels, the visible part of the PV system, and does not include all the other hardware, often summarized as balance of system (BOS). Moreover, PV systems convert light directly into electricity and shouldn't be confused with other technologies, such as concentrated solar power or solar thermal, used for heating and cooling.

PV systems range from small, rooftop-mounted or building-integrated systems with capacities from a few to several tens of kilowatts, to large utility-scale power stations of hundreds of megawatts. Nowadays, most PV systems are grid-connected, while off-grid or stand-alone systems only account for a small portion of the market.

Operating silently and without any moving parts or environmental emissions, PV systems have developed from being niche market applications into a mature technology used for mainstream electricity generation. A rooftop system recoups the invested energy for its manufacturing and installation within 0.7 to 2 years and produces about 95 percent of net clean renewable energy over a 30-year service lifetime.

Due to the exponential growth of photovoltaics, prices for PV systems have rapidly declined in recent years. However, they vary by market and the size of the system. In 2014, prices for residential 5-kilowatt systems in the United States were around $3.29 per watt, while in the highly penetrated German market, prices for rooftop systems of up to 100 kW declined to €1.24 per watt. Nowadays, solar PV modules account for less than half of the system's overall cost, leaving the rest to the remaining BOS-components and to soft costs, which include customer acquisition, permitting, inspection and interconnection, installation labor and financing costs.

Modern System

Overview

A photovoltaic system converts the sun's radiation into usable electricity. It comprises the solar array and the balance of system components. PV systems can be categorized by various aspects, such as, grid-connected vs. stand alone systems, building-integrated vs. rack-mounted systems, residential vs. utility systems, distributed vs. centralized systems, rooftop vs. ground-mounted systems, tracking vs. fixed-tilt systems, and new constructed vs. retrofitted systems. Other distinctions may include, systems with microinverters vs. central inverter, systems using crystalline silicon vs. thin-film technology, and systems with modules from Chinese vs. European and U.S.-manufacturers.

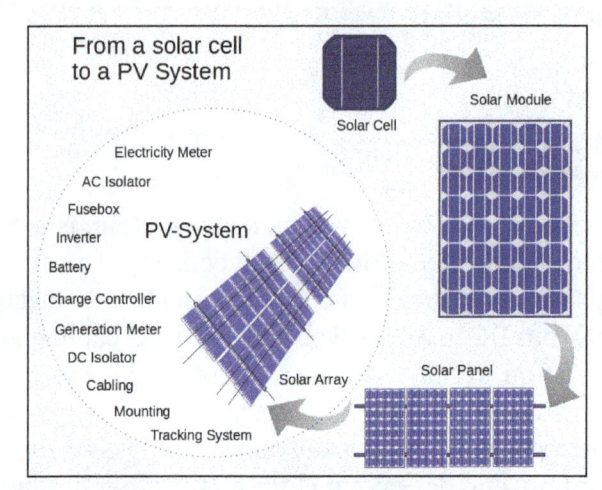

Diagram of the possible components of a photovoltaic system.

About 99 percent of all European and 90 percent of all U.S. solar power systems are connected to the electrical grid, while off-grid systems are somewhat more common in Australia and South Korea. PV systems rarely use battery storage. This may change soon, as government incentives for distributed energy storage are being implemented and investments in storage solutions are gradually becoming economically viable for small systems. A solar array of a typical residential PV system is rack-mounted on the roof, rather than integrated into the roof or facade of the building, as this is significantly more expensive. Utility-scale solar power stations are ground-mounted, with fixed tilted solar panels rather than using expensive tracking devices. Crystalline silicon is the predominant material used in 90 percent of worldwide produced solar modules, while rival thin-film has lost market-share in recent years. About 70 percent of all solar cells and modules are produced in China and Taiwan, leaving only 5 percent to European and US-manufacturers. The installed capacity for both, small rooftop systems and large solar power stations is growing rapidly and in equal parts, although there is a notable trend towards utility-scale systems, as the focus on new installations is shifting away from Europe to sunnier regions, such as the Sunbelt in the U.S., which are less opposed to ground-mounted solar farms and cost-effectiveness is more emphasized by investors.

Driven by advances in technology and increases in manufacturing scale and sophistication, the cost of photovoltaics is declining continuously. There are several million PV systems distributed all over the world, mostly in Europe, with 1.4 million systems in Germany alone– as well as North America with 440,000 systems in the United States, The energy conversion efficiency of a conventional solar module increased from 15 to 20 percent over the last 10 years and a PV system recoups the energy needed for its manufacture in about 2 years. In exceptionally irradiated locations, or when thin-film technology is used, the so-called energy payback time decreases to one year or less. Net metering and financial incentives, such as preferential feed-in tariffs for solar-gen-erated electricity, have also greatly supported installations of PV systems in many countries. The levelised cost of electricity from large-scale PV systems has become competitive with conventional electricity sources in an expanding list of geographic regions, and grid parity has been achieved in about 30 different countries.

As of 2015, the fast-growing global PV market is rapidly approaching the 200 GW mark – about 40 times the installed capacity of 2006. Photovoltaic systems currently contribute about 1 percent to worldwide electricity generation. Top installers of PV systems in terms of capacity are currently China, Japan and the United States, while half of the world's capacity is installed in Europe, with Germany and Italy supplying 7% to 8% of their respective domestic electricity consumption with solar PV. The International Energy Agency expects solar power to become the world's largest source of electricity by 2050, with solar photovoltaics and concentrated solar thermal contributing 16% and 11% to the global demand, respectively.

Grid-connection

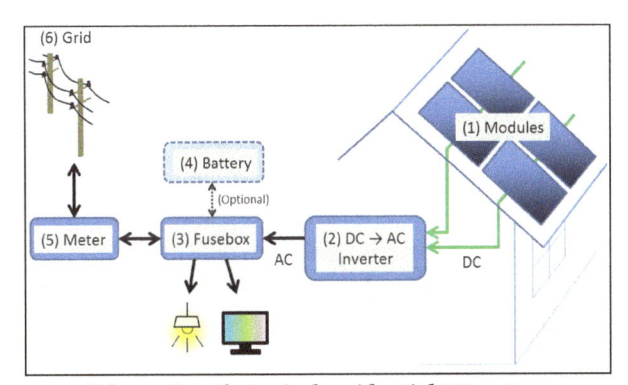

Schematics of a typical residential PV system.

A grid connected system is connected to a larger independent grid (typically the public electricity grid) and feeds energy directly into the grid. This energy may be shared by a residential or commercial building before or after the revenue measurement point. The difference being whether the credited energy production is calculated independently of the customer's energy consumption (feed-in tariff) or only on the difference of energy (net metering). Grid connected systems vary in size from residential (2–10 kW_p) to solar power stations (up to 10s of MW_p). This is a form of decentralized electricity generation. The feeding of electricity into the grid requires the transformation of DC into AC by a special, synchronising grid-tie inverter. In kilowatt-sized installations the DC side system voltage is as high as permitted (typically 1000V except US residential 600 V) to limit ohmic losses. Most modules (60 or 72 crystalline silicon cells) generate 160 W to 300 W at 36 volts. It is sometimes necessary or desirable to connect the modules partially in parallel rather than all in series. One set of modules connected in series is known as a 'string'.

Scale of System

Photovoltaic systems are generally categorized into three distinct market segments: residential rooftop, commercial rooftop, and ground-mount utility-scale systems. Their capacities range from a few kilowatts to hundreds of megawatts. A typical residential system is around 10 kilowatts and mounted on a sloped roof, while commercial systems may reach a megawatt-scale and are generally installed on low-slope or even flat roofs. Although rooftop mounted systems are small and display a higher cost per watt than large utility-scale installations, they account for the largest share in the market. There is, however, a growing trend towards bigger utility-scale power plants, especially in the "sunbelt" region of the planet.

Utility-scale

Perovo Solar Park in Ukraine.

Large utility-scale solar parks or farms are power stations and capable of providing an energy supply to large numbers of consumers. Generated electricity is fed into the transmission grid powered by central generation plants (grid-connected or grid-tied plant), or combined with one, or many, domestic electricity generators to feed into a small electrical grid (hybrid plant). In rare cases generated electricity is stored or used directly by island/standalone plant. PV systems are generally designed in order to ensure the highest energy yield for a given investment. Some large photovoltaic power stations such as Solar Star, Waldpolenz Solar Park and Topaz Solar Farm cover tens or hundreds of hectares and have power outputs up to hundreds of megawatts.

Rooftop, mobile, and portable

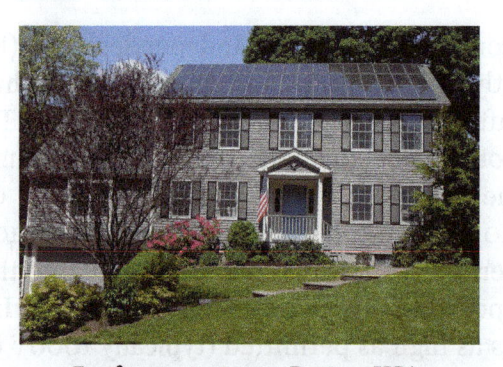

Rooftop system near Boston, USA.

A small PV system is capable of providing enough AC electricity to power a single home, or even an isolated device in the form of AC or DC electric. For example, military and civilian Earth observation satellites, street lights, construction and traffic signs, electric cars, solar-powered tents, and electric aircraft may contain integrated photovoltaic systems to provide a primary or auxiliary power source in the form of AC or DC power, depending on the design and power demands. In 2013, rooftop systems accounted for 60 percent of worldwide installations. However, there is a trend away from rooftop and towards utility-scale PV systems, as the focus of new PV installations is also shifting from Europe to countries in the sunbelt region of the planet where opposition to ground-mounted solar farms is less accentuated.

Portable and mobile PV systems provide electrical power independent of utility connections, for "off the grid" operation. Such systems are so commonly used on recreational vehicles and boats that there are retailers specializing in these applications and products specifically targeted to them. Since recreational vehicles (RV) normally carry batteries and operate lighting and other systems on nominally 12-volt DC power, RV PV systems normally operate in a voltage range chosen to charge 12-volt batteries directly, and addition of a PV system requires only panels, a charge controller, and wiring.

Building-integrated

BAPV wall near Barcelona, Spain.

In urban and suburban areas, photovoltaic arrays are commonly used on rooftops to supplement power use; often the building will have a connection to the power grid, in which case the energy produced by the PV array can be sold back to the utility in some sort of net metering agreement. Some utilities, such as Solvay Electric in Solvay, NY, use the rooftops of commercial customers and telephone poles to support their use of PV panels. Solar trees are arrays that, as the name implies, mimic the look of trees, provide shade, and at night can function as street lights.

Performance

Uncertainties in revenue over time relate mostly to the evaluation of the solar resource and to the performance of the system itself. In the best of cases, uncertainties are typically 4% for year-to-year climate variability, 5% for solar resource estimation (in a horizontal plane), 3% for estimation of irradiation in the plane of the array, 3% for power rating of modules, 2% for losses due to dirt and soiling, 1.5% for losses due to snow, and 5% for other sources of error. Identifying and reacting to manageable losses is critical for revenue and O&M efficiency. Monitoring of array performance may be part of contractual agreements between the array owner, the builder, and the utility purchasing the energy produced. Recently, a method to create "synthetic days" using readily available weather data and verification using the Open Solar Outdoors Test Field make it possible to predict photovoltaic systems performance with high degrees of accuracy. This method can be used to then determine loss mechanisms on a local scale - such as those from snow or the effects of surface coatings (e.g. hydrophobic or hydrophilic) on soiling or snow losses. (Although in heavy snow environments with severe ground interference can result in annual losses from snow of 30%.) Access to the Internet has allowed a further improvement in energy monitoring and communication. Dedicated systems are available from a number of vendors. For solar PV system that use microinverters (panel-level

DC to AC conversion), module power data is automatically provided. Some systems allow setting performance alerts that trigger phone/email/text warnings when limits are reached. These solutions provide data for the system owner and the installer. Installers are able to remotely monitor multiple installations, and see at-a-glance the status of their entire installed base.

Components

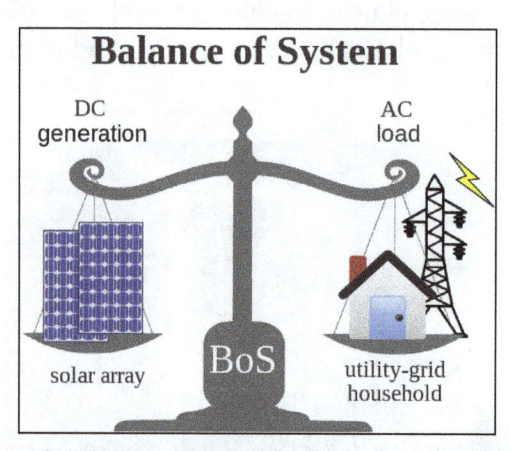

The balance of system components of a PV system (BOS) balance the power-generating subsystem of the solar array (left side) with the power-using side of the AC-household devices and the utility grid (right side).

A photovoltaic system for residential, commercial, or industrial energy supply consists of the solar array and a number of components often summarized as the balance of system (BOS). The term originates from the fact that some BOS-components are balancing the power-generating subsystem of the solar array with the power-using side, the load. BOS-components include power-conditioning equipment and structures for mounting, typically one or more DC to AC power converters, also known as inverters, an energy storage device, a racking system that supports the solar array, electrical wiring and interconnections, and mounting for other components.

Optionally, a balance of system may include any or all of the following: renewable energy credit revenue-grade meter, maximum power point tracker (MPPT), battery system and charger, GPS solar tracker, energy management software, solar irradiance sensors, anemometer, or task-specific accessories designed to meet specialized requirements for a system owner. In addition, a CPV system requires optical lenses or mirrors and sometimes a cooling system.

The terms *"solar array"* and *"PV system"* are often used interchangeably, despite the fact that the solar array does not encompass the entire system. Moreover, *"solar panel"* is often used as a synonym for *"solar module"*, although a panel consists of a string of several modules. The term *"solar system"* is also an often used misnomer for a PV system.

Solar Array

Conventional c-Si solar cells, normally wired in series, are encapsulated in a solar module to protect them from the weather. The module consists of a tempered glass as cover, a soft and flexible encapsulant, a rear backsheet made of a weathering and fire-resistant material and an aluminium frame around the outer edge. Electrically connected and mounted on a supporting structure, solar modules build a string of modules, often called solar panel. A solar array consists of one or many such panels. A photovoltaic array, or solar array, is a linked collection of solar panels. The power that one module can produce is seldom enough to meet requirements of a home or a business, so the modules are linked together to form an array. Most PV arrays use an inverter to convert the DC power produced by the modules into alternating current that can power lights, motors, and other loads. The modules in a PV array are usually first connected in series to obtain the desired voltage; the individual strings are then connected in parallel to allow the system to produce more current. Solar panels are typically measured under STC (standard test conditions) or PTC (PVUSA test conditions), in watts. Typical panel ratings range from less than 100 watts to over 400 watts. The array rating consists of a summation of the panel ratings, in watts, kilowatts, or megawatts.

Module and Efficiency

A typical "150 watt" PV module is about a square meter in size. Such a module may be expected to produce 0.75 kilowatt-hour (kWh) every day, on average, after taking into account the weather and the latitude, for an insolation of 5 sun hours/day. In the last 10 years, the efficiency of average commercial wafer-based crystalline silicon modules increased from about 12% to 16% and CdTe module efficiency increased from 9% to 13% during same period. Module output and life degraded by increased temperature. Allowing ambient air to flow over, and if possible behind, PV modules reduces this problem. Effective module lives are typically 25 years or more. The payback period for an investment in a PV solar installation varies greatly and is typically less useful than a calculation of return on investment. While it is typically calculated to be between 10 and 20 years, the financial payback period can be far shorter with incentives.

Fixed tilt solar array in of crystalline silicon panels in Canterbury, New Hampshire, United States.

Solar array of a solar farm with a few thousand solar modules on the island of Majorca, Spain.

Due to the low voltage of an individual solar cell (typically ca. 0.5V), several cells are wired in series in the manufacture of a "laminate". The laminate is assembled into a protective weatherproof enclosure, thus making a photovoltaic module or solar panel. Modules may then be strung together into a photovoltaic array. In 2012, solar panels avail-able for consumers can have an efficiency of up to about 17%, while commercially available panels can go as far as 27%. It has been recorded that a group from the The Fraunhofer Institute for Solar Energy Systems have created a cell that can reach 44.7% efficiency, which makes scientists' hopes of reaching the 50% efficiency threshold a lot more feasible.

Shading and Dirt

Photovoltaic cell electrical output is extremely sensitive to shading. The effects of this shading are well known. When even a small portion of a cell, module, or array is shaded, while the remainder is in sunlight, the output falls dramatically due to internal 'short-circuiting' (the electrons reversing course through the shaded portion of the p-n junction). If the current drawn from the series string of cells is no greater than the current that can be produced by the shaded cell, the current (and so power) developed by the string is limited. If enough voltage is available from the rest of the cells in a string, current will be forced through the cell by breaking down the junction in the shaded portion. This breakdown voltage in common cells is between 10 and 30 volts. Instead of adding to the power produced by the panel, the shaded cell absorbs power, turning it into heat. Since the reverse voltage of a shaded cell is much greater than the forward voltage of an illuminated cell, one shaded cell can absorb the power of many other cells in the string, disproportionately affecting panel output. For example, a shaded cell may drop 8 volts, instead of adding 0.5 volts, at a particular current level, thereby absorbing the power produced by 16 other cells. It is, thus important that a PV installation is not shaded by trees or other obstructions.

Several methods have been developed to determine shading losses from trees to PV systems over both large regions using LiDAR, but also at an individual system level using sketchup. Most modules have bypass diodes between each cell or string of cells that minimize the effects of shading and only lose the power of the shaded portion of the array. The main job of the bypass diode is to eliminate hot spots that form on cells that can cause further damage to the array, and cause fires. Sunlight can be absorbed by dust, snow, or other impurities at the surface of the module. This can reduce the light that strikes the cells. In general these losses aggregated over the year are small even for locations in Canada. Maintaining a clean module surface will increase output performance over the life of the module. Google found that cleaning the flat mounted solar panels after 15 months increased their output by almost 100%, but that the 5% tilted arrays were adequately cleaned by rainwater.

Insolation and Energy

Solar insolation is made up of direct, diffuse, and reflected radiation. The absorption factor of a PV cell is defined as the fraction of incident solar irradiance that is absorbed by the cell. At high noon on a cloudless day at the equator, the power of the sun is about 1 kW/m², on the Earth's surface, to a plane that is perpendicular to the sun's rays. As such, PV arrays can track the sun through each day to greatly enhance energy collection. However, tracking devices add cost, and require maintenance, so it is more common for PV arrays to have fixed mounts that tilt the array and face solar noon (approximately due south in the Northern Hemisphere or due north in the Southern Hemisphere). The tilt angle, from horizontal, can be varied for season, but if fixed, should be set to give optimal array output during the peak electrical demand portion of a typical year for a stand-alone system. This optimal module tilt angle is not necessarily identical to the tilt angle for maximum annual array energy output. The optimization of the a photovoltaic system for a specific environment can be complicated as issues of solar flux, soiling, and snow losses should be taken into effect. In addition, recent work has shown that spectral effects can play a role in optimal photovoltaic material selection. For example, the spectral albedo can play a significant role in output depending on the surface around the photovoltaic system and the type of solar cell material. For the weather and latitudes of the United States and Europe, typical insolation ranges from 4 kWh/m^2/day in northern climes to 6.5 kWh/m^2/day in the sunniest regions. A photovoltaic installation in the southern latitudes of Europe or the United States may expect to produce 1 kWh/m^2/day. A typical 1 kW photovoltaic installation in Australia or the southern latitudes of Europe or United States, may produce 3.5–5 kWh per day, dependent on location, orientation, tilt, insolation and other factors. In the Sahara desert, with less cloud cover and a better solar angle, one could ideally obtain closer to 8.3 kWh/m^2/day provided the nearly ever present wind would not blow sand onto the units. The area of the Sahara desert is over 9 million km^2. 90,600 km^2, or about 1%, could generate as much electricity as all of the world's power plants combined.

Mounting

A 23-year-old, ground mounted PV system from the 1980s on a North Frisian Island, Germany.
The modules conversion efficiency was only 12%.

Modules are assembled into arrays on some kind of mounting system, which may be classified as ground mount, roof mount or pole mount. For solar parks a large rack is mounted on the ground, and the modules mounted on the rack. For buildings, many different racks have been devised for pitched roofs. For flat roofs, racks, bins and building integrated solutions are used. Solar panel racks mounted on top of poles can be stationary or moving, see Trackers below. Side-of-pole mounts are suitable for situations where a pole has something else mounted at its top, such as a light fixture or an antenna. Pole mounting raises what would otherwise be a ground mounted array above weed shadows and livestock, and may satisfy electrical code requirements regarding inaccessibility of exposed wiring. Pole mounted panels are open to more cooling air on their underside, which increases performance. A multiplicity of pole top racks can be formed into a parking carport or other shade structure. A rack which does not follow the sun from left to right may allow seasonal adjustment up or down.

Cabling

Due to their outdoor usage, solar cables are specifically designed to be resistant against UV radiation and extremely high temperature fluctuations and are generally unaffected by the weather. A number of standards specify the usage of electrical wiring in PV systems, such as the IEC 60364 by the International Electrotechnical Commission, in section 712 *"Solar photovoltaic (PV) power supply systems"*, the British Standard BS 7671, incorporating regulations relating to microgeneration and photovoltaic systems, and the US UL4703 standard, in subject 4703 *"Photovoltaic Wire"*.

Tracker

A 1998 model of a passive solar tracker, viewed from underneath.

A solar tracking system tilts a solar panel throughout the day. Depending on the type of tracking system, the panel is either aimed directly at the sun or the brightest area of a partly clouded sky. Trackers greatly enhance early morning and late afternoon performance, increasing the total amount of power produced by a system by about 20–25% for a single axis tracker and about 30% or more for a dual axis tracker, depending on latitude. Trackers are effective in regions that receive a large portion of sunlight directly. In diffuse light (i.e. under cloud or fog), tracking has little or no value. Because most concentrated photovoltaics systems are very sensitive to the sunlight's angle, tracking systems allow them to produce useful power for more than a brief period each day. Tracking systems improve performance for two main reasons. First, when a solar panel is perpendicular to the sunlight, it receives more light on its surface than if it were angled. Second, direct light is used more efficiently than angled light. Special Anti-reflective coatings can improve solar panel efficiency for direct and angled light, somewhat reducing the benefit of tracking.

Trackers and sensors to optimise the performance are often seen as optional, but tracking systems can increase viable output by up to 45%. PV arrays that approach or exceed one megawatt often use solar trackers. Accounting for clouds, and the fact that most of the world is not on the equator, and that the sun sets in the evening, the correct measure of solar power is insolation – the average number of kilowatt-hours per square meter per day. For the weather and latitudes of the United States and Europe, typical insolation ranges from 2.26 kWh/m²/day in northern climes to 5.61 kWh/m²/day in the sunniest regions.

For large systems, the energy gained by using tracking systems can outweigh the added complexity (trackers can increase efficiency by 30% or more). For very large systems, the added maintenance of tracking is a substantial detriment. Tracking is not required for flat panel and low-concentration photovoltaic systems. For high-concentration photovoltaic systems, dual axis tracking is a necessity. Pricing trends affect the balance between adding more stationary solar panels versus having fewer panels that track. When solar panel prices drop, trackers become a less attractive option.

Inverter

Central inverter with AC and DC disconnects (on the side), monitoring gateway, transformer isolation and interactive LCD.

Systems designed to deliver alternating current (AC), such as grid-connected applications need an inverter to convert the direct current (DC) from the solar modules to AC. Grid connected inverters must supply AC electricity in sinusoidal form, synchronized to the grid frequency, limit feed in voltage to no higher than the grid voltage and disconnect from the grid if the grid voltage is turned off. Islanding inverters need only produce regulated voltages and frequencies in a sinusoidal waveshape as no synchronisation or co-ordination with grid supplies is required.

String inverter (left), generation meter, and AC disconnect (right). A modern 2013 installation in Vermont, United States.

A solar inverter may connect to a string of solar panels. In some installations a solar micro-inverter is connected at each solar panel. For safety reasons a circuit breaker is provided both on the AC and DC side to enable maintenance. AC output may be connected through an electricity meter into the public grid. The number of modules in the system determines the total DC watts capable of being generated by the solar array; however, the inverter ultimately governs the amount of AC watts that can be distributed for consumption. For example, a PV system comprising 11 kilowatts DC (kW_{DC}) worth of PV modules, paired with one 10-kilowatt AC (kW_{AC}) inverter, will be limited to the inverter's output of 10 kW. As of 2014, conversion efficiency for state-of-the-art converters reached more than 98 percent. While string inverters are used in residential to medium-sized commercial PV systems, central inverters cover the large commercial and utility-scale market. Market-share for central and string inverters are about 50 percent and 48 percent, respectively, leaving less than 2 percent to micro-inverters.

Maximum power point tracking (MPPT) is a technique that grid connected inverters use to get the maximum possible power from the photovoltaic array. In order to do so, the inverter's MPPT system digitally samples the solar array's ever changing power output and applies the proper resistance to find the optimal *maximum power point*.

Anti-islanding is a protection mechanism that immediately shuts down the inverter preventing it from generating AC power when the connection to the load no longer exists. This happens, for example, in the case of a blackout. Without this protection, the supply line would become an "island" with power surrounded by a "sea" of unpowered lines, as the solar array continues to deliver DC power during the power outage. Islanding is a hazard to utility workers, who may not realize that an AC circuit is still powered, and it may prevent automatic re-connection of devices.

Inverter/Converter Market in 2014				
Type	**Power**	**Efficiency**[a]	**Market Share**[b]	**Remarks**
String inverter	up to 100 kW_p [c]	98%	50%	Cost[b] €0.15 per watt-peak. Easy to replace.
Central inverter	above 100 kW_p	98.5%	48%	€0.10 per watt-peak. High reliability. Often sold along with a service contract.
Micro-inverter	module power range	90%–95%	1.5%	€0.40 per watt-peak. Ease of replacement concerns.
DC/DC converter Power optimizer	module power range	98.8%	n.a.	€0.40 per watt-peak. Ease of replacement concerns. Inverter is still needed. About 0.75 GW_p installed in 2013.
Source: *data by IHS 2014, remarks by Fraunhofer ISE 2014, from: Photovoltaics Report, updated as per 8 September 2014, p. 35, PDF* Notes: [a]*best efficiencies displayed,* [b]*market-share and cost per watt are estimated,* [c]kW_p *= kilowatt-peak*				

Battery

Although still expensive, PV systems increasingly use rechargeable batteries to store a surplus to be later used at night. Batteries used for grid-storage also stabilize the electrical grid by leveling out peak loads, and play an important role in a smart grid, as they can charge during periods of low demand and feed their stored energy into the grid when demand is high.

Common battery technologies used in today's PV systems include the valve regulated lead-acid battery– a modified version of the conventional lead–acid battery, nickel–cadmium and lithium-ion batteries. Compared to the other types, lead-acid batteries have a shorter lifetime and lower energy density. However, due to their high reliability, low self discharge as well as low investment and maintenance costs, they are currently the predominant technology used in small-scale, residential

PV systems, as lithium-ion batteries are still being developed and about 3.5 times as expensive as lead-acid batteries. Furthermore, as storage devices for PV systems are stationary, the lower energy and power density and therefore higher weight of lead-acid batteries are not as critical as, for example, in electric transportation. Other rechargeable batteries that are considered for distributed PV systems include sodium–sulfur and vanadium redox batteries, two prominent types of a molten salt and a low battery, respectively. In 2015, Tesla motors launched the Powerwall, a rechargeable lithiumion battery with the aim to revolutionize energy consumption.

PV systems with an integrated battery solution also need a charge controller, as the varying voltage and current from the solar array requires constant adjustment to prevent damage from overcharging. Basic charge controllers may simply turn the PV panels on and off, or may meter out pulses of energy as needed, a strategy called PWM or pulse-width modulation. More advanced charge controllers will incorporate MPPT logic into their battery charging algorithms. Charge controllers may also divert energy to some purpose other than battery charging. Rather than simply shut off the free PV energy when not needed, a user may choose to heat air or water once the battery is full.

Monitoring and Metering

The metering must be able to accumulate energy units in both directions or two meters must be used. Many meters accumulate bidirectionally, some systems use two meters, but a unidirectional meter (with detent) will not accumulate energy from any resultant feed into the grid. In some countries, for installations over 30 kW_p a frequency and a voltage monitor with disconnection of all phases is required. This is done where more solar power is being generated than can be accommodated by the utility, and the excess can not either be exported or stored. Grid operators historically have needed to provide transmission lines and generation capacity. Now they need to also provide storage. This is normally hydro-storage, but other means of storage are used. Initially storage was used so that baseload generators could operate at full output. With variable renewable energy, storage is needed to allow power generation whenever it is available, and consumption whenever it is needed.

A Canadian electricity meter.

The two variables a grid operator have are storing electricity for *when* it is needed, or transmitting it to *where* it is needed. If both of those fail, installations over 30kWp can automatically shut down, although in practice all inverters maintain voltage regulation and stop supplying power if the load is inadequate. Grid operators have the option of curtailing excess generation from large systems, although this is more commonly done with wind power than solar power, and results in a substantial loss of revenue. Three-phase inverters have the unique option of supplying reactive power which can be advantageous in matching load requirements.

Photovoltaic systems need to be monitored to detect breakdown and optimize their operation. There are several *photovoltaic monitoring* strategies depending on the output of the installation and its nature. Monitoring can be performed on site or remotely. It can measure production only, retrieve all the data from the inverter or retrieve all of the data from the communicating equipment (probes, meters, etc.). Monitoring tools can be dedicated to supervision only or offer additional functions. Individual inverters and battery charge controllers may include monitoring using manufacturer specific protocols and software. Energy metering of an inverter may be of limited accuracy and not suitable for revenue metering purposes. A third-party data acquisition system can monitor multiple inverters, using the inverter manufacturer's protocols, and also acquire weather-related information. Independent smart meters may measure the total energy production of a PV array system. Separate measures such as satellite image analysis or a solar radiation meter (a pyranometer) can be used to estimate total insolation for comparison. Data collected from a monitoring system can be displayed remotely over the World Wide Web, such as OSOTF.

Other Systems

This section includes systems that are either highly specialized and uncommon or still an emerging new technology with limited significance. However, standalone or off-grid systems take a special place. They were the most common type of systems during the 1980s and 1990s, when PV technology was still very expensive and a pure niche market of small scale applications. Only in places where no electrical grid was available, they were economically viable. Although new stand-alone systems are still being deployed all around the world, their contribution to the overall installed photovoltaic capacity is decreasing. In Europe, off-grid systems account for 1 percent of installed capacity. In the United States, they account for about 10 percent. Off-grid systems are still common in Australia and South Korea, and in many developing countries.

CPV

Concentrator photovoltaic (CPV) in Catalonia, Spain.

Concentrator photovoltaics (CPV) and *high concentrator photovoltaic* (HCPV) systems use optical lenses or curved mirrors to concentrate sunlight onto small but highly efficient solar cells. Besides concentrating optics, CPV systems sometime use solar trackers and cooling systems and are more expensive.

Especially HCPV systems are best suited in location with high solar irradiance, concentrating sunlight up to 400 times or more, with efficiencies of 24–28 percent, exceeding those of regular systems. Various designs of CPV and HCPV systems are commercially available but not very common. However, ongoing research and development is taking place.

CPV is often confused with CSP (concentrated solar power) that does not use photovoltaics. Both technologies favor locations that receive much sunlight and are directly competing with each other.

Hybrid

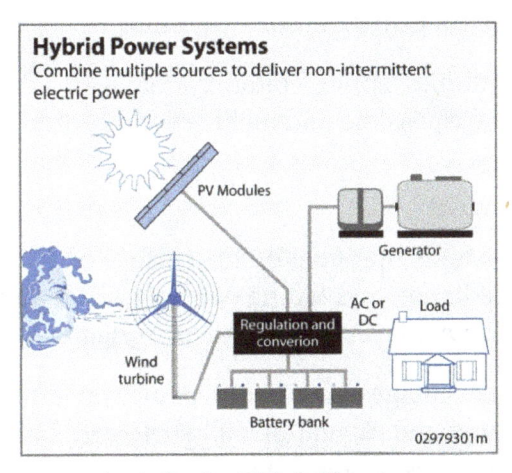

A wind-solar PV hybrid system.

A hybrid system combines PV with other forms of generation, usually a diesel generator. Biogas is also used. The other form of generation may be a type able to modulate power output as a function of demand. However more than one renewable form of energy may be used e.g. wind. The photovoltaic power generation serves to reduce the consumption of non renewable fuel. Hybrid systems are most often found on islands. Pellworm island in Germany and Kythnos island in Greece are notable examples (both are combined with wind). The Kythnos plant has reduced diesel consumption by 11.2%.

In 2015, a case-study conducted in seven countries concluded that in all cases generating costs can be reduced by hybridising mini-grids and isolated grids. However, financing costs for such hybrids are crucial and largely depend on the ownership structure of the power plant. While cost reductions for state-owned utilities can be significant, the study also identified economic benefits to be insignificant or even negative for non-public utilities, such as independent power producers.

There has also been recent work showing that the PV penetration limit can be increased by deploying a distributed network of PV+CHP hybrid systems in the U.S. The temporal distribution of solar flux, electrical and heating requirements for representative U.S. single family residences were analyzed and the results clearly show that hybridizing CHP with PV can enable additional PV deployment above what is possible with a conventional centralized electric generation system. This theory was reconfirmed with numerical simulations using per second solar flux data to determine that the necessary battery backup to provide for such a hybrid system is possible with relatively small and inexpensive battery systems. In addition, large PV+CHP systems are possible for institutional buildings, which again provide back up for intermittent PV and reduce CHP runtime.

- PVT system (hybrid PV/T), also known as *photovoltaic thermal hybrid solar collectors* convert solar radiation into thermal and electrical energy. Such a system combines a solar (PV) module with a solar thermal collector in an complementary way.

- CPVT system. A *concentrated photovoltaic thermal hybrid* (CPVT) system is similar to a PVT system. It uses concentrated photovoltaics (CPV) instead of conventional PV technology, and combines it with a solar thermal collector.

- CPV/CSP system. A novel solar CPV/CSP hybrid system has been proposed recently, combining concentrator photovoltaics with the non-PV technology of concentrated solar power (CSP), or also known as concentrated solar thermal.

- PV diesel system. It combines a photovoltaic system with a diesel generator. Combinations with other renewables are possible and include wind turbines.

Floating Solar Arrays

Floating solar arrays are PV systems that float on the surface of drinking water reservoirs, quarry lakes, irrigation canals or remediation and tailing ponds. A small number of such systems exist in France, India, Japan, South Korea, the United Kingdom, Singapore and the United States.

The systems are said to have advantages over photovoltaics on land. The cost of land is more expensive, and there are fewer rules and regulations for structures built on bodies of water not used for recreation. Unlike most land-based solar plants, floating arrays can be unobtrusive because they are hidden from public view. They achieve higher efficiencies than PV panels on land, because water cools the panels. The panels have a special coating to prevent rust or corrosion.

In May 2008, the Far Niente Winery in Oakville, California, pioneered the world's first floatovoltaic system by installing 994 solar PV modules with a total capacity of 477 kW onto 130 pontoons and floating them on the winery's irrigation pond. The primary benefit of such a system is that it avoids the need to sacrifice valuable land area that could be used for another purpose. In the case of the Far Niente Winery, it saved three-quarters of an acre that would have been required for a land-based system. Another benefit of a floatovoltaic system is that the panels are kept at a cooler temperature than they would be on land, leading to a higher efficiency of solar energy conversion. The floating PV array also reduces the amount of water lost through evaporation and inhibits the growth of algae.

Utility-scale floating PV farms are starting to be built. The multinational electronics and ceramics manufacturer Kyocera will develop the world's largest, a 13.4 MW farm on the reservoir above Yamakura Dam in Chiba Prefecture using 50,000 solar panels. Salt-water resistant floating farms are also being considered for ocean use, with experiments in Thailand. The largest so far announced floatovoltaic project is a 350 MW power station in the Amazon region of Brazil.

Direct Current Grid

DC grids are found in electric powered transport: railways trams and trolleybuses. A few pilot plants for such applications have been built, such as the tram depots in Hannover Leinhausen, using photovoltaic contributors and Geneva (Bachet de Pesay). The 150 kW$_p$ Geneva site feeds 600V DC directly into the tram/trolleybus electricity network whereas before it provided about 15% of the electricity at its opening in 1999.

Standalone

A stand-alone or off-grid system is not connected to the electrical grid. Standalone systems vary widely in size and application from wristwatches or calculators to remote buildings or spacecraft. If the load is to be supplied independently of solar insolation, the generated power is stored and buffered with a battery. In non-portable applications where weight is not an issue, such as in buildings, lead acid batteries are most commonly used for their low cost and tolerance for abuse.

An isolated mountain hut in Catalonia, Spain.

Solar parking meter in Edinburgh, Scotland.

A charge controller may be incorporated in the system to avoid battery damage by excessive charging or discharging. It may also help to optimize production from the solar array using a maximum power point tracking technique (MPPT). However, in simple PV systems where the PV module voltage is matched to the battery voltage, the use of MPPT electronics is generally considered unnecessary, since the battery voltage is stable enough to provide near-maximum power collection from the PV module. In small devices (e.g. calculators, parking meters) only direct current (DC) is consumed. In larger systems (e.g. buildings, remote water pumps) AC is usually required. To convert the DC from the modules or batteries into AC, an inverter is used.

In agricultural settings, the array may be used to directly power DC pumps, without the need for an inverter. In remote settings such as mountainous areas, islands, or other places where a power grid is unavailable, solar arrays can be used as the sole source of electricity, usually by charging a storage battery. Stand-alone systems closely relate to microgeneration and distributed generation.

- Pico PV systems:

The smallest, often portable photovoltaic systems are called pico solar PV systems, or pico solar. They mostly combine a rechargeable battery and charge controller, with a very small PV panel. The panel's nominal capacity is just a few watt-peak (1–10 W_p) and its area less than a tenth of a square meter, or one square foot, in size. A large range of different applications can be solar powered such as music players, fans, portable lamps, security lights, solar lighting kits, solar lanterns and street light (see below), phone chargers, radios, or even small, seven-inch LCD televisions, that run on less than ten watts. As it is the case for power generation from pico hydro, pico PV systems are useful in small, rural communities that require only a small amount of electricity. Since the efficiency of many appliances have improved considerably, in particular due to the usage of LED lights and efficient rechargeable batteries, pico solar has become an affordable alternative, especially in the developing world. The metric prefix pico- stands for a trillionth to indicate the smallness of the system's electric power.

- Solar street lights:

Solar street lights raised light sources which are powered by photovoltaic panels generally mounted on the lighting structure. The solar array of such off-grid PV system charges a rechargeable battery, which powers a fluorescent or LED lamp during the night. Solar street lights are stand-alone power systems, and have the advantage of savings on trenching, landscaping, and maintenance costs, as well as on the electric bills, despite their higher initial cost compared to conventional street lighting. They are designed with sufficiently large batteries to ensure operation for at least a week and even in the worst situation, they are expected to dim only slightly.

- Telecommunication and signaling:

Solar PV power is ideally suited for telecommunication applications such as local telephone exchange, radio and TV broadcasting, microwave and other forms of electronic communication links. This is because, in most telecommunication application, storage batteries are already in use and the electrical system is basically DC. In hilly and mountainous terrain, radio and TV signals may not reach as they get blocked or reflected back due to undulating terrain. At these locations, low power transmitters are installed to receive and retransmit the signal for local population.

- Solar vehicles:

Solar vehicle, whether ground, water, air or space vehicles may obtain some or all of the energy required for their operation from the sun. Surface vehicles generally require higher power levels than can be sustained by a practically sized solar array, so a battery assists in meeting peak power demand, and the solar array recharges it. Space vehicles have successfully used solar photovoltaic systems for years of operation, eliminating the weight of fuel or primary batteries.

- Solar pumps:

One of the most cost effective solar applications is a solar powered pump, as it is far cheaper to purchase a solar panel than it is to run power lines. They often meet a need for water beyond the reach of power lines, taking the place of a windmill or windpump. One common application is the filling of livestock watering tanks, so that grazing cattle may drink. Another is the refilling of drinking water storage tanks on remote or self-sufficient homes.

- Spacecraft:

Solar panels on spacecraft have been one of the first applications of photovoltaics since the launch of Vanguard 1 in 1958, the first satellite to use solar cells. Contrary to Sputnik, the first artificial satellite to orbit the planet, that ran out of batteries within 21 days due to the lack of solar-power, most modern communications satellites and space probes in the inner solar system rely on the use of solar panels to derive electricity from sunlight.

- Do it yourself community:

With a growing interest in environmentally friendly green energy, an increasing number of hobbyists in the DIY-community have endeavored to build their own solar PV systems from kits or partly DIY. Usually, the DIY-community uses inexpensive or high efficiency systems (such as those with solar tracking) to generate their own power. As a result, the DIY-systems often end up cheaper than their commercial counterparts. Often, the system is also hooked up into the regular power grid, using net metering instead of a battery for backup. These systems usually generate power amount of ~2 kW or less. Through the internet, the community is now able to obtain plans to (partly) construct the system and there is a growing trend toward building them for domestic requirements.

Gallery of Standalone Systems

 Artist's concept of the Juno spacecraft orbiting Jupiter.	 Solar powered electric fence, in Harwood Northumberland, UK.	 Powering a Yurt in Mongolia.
 A small solar water pump system.	 Solar car. The Japanese winner of 2009 World Solar Challenge in Australia.	 A solar sewage treatment plant in Santuari de Lluc, Spain.

Costs and Economy

Median installed system prices for residential PV Systems in Japan, Germany and the United States ($/W).

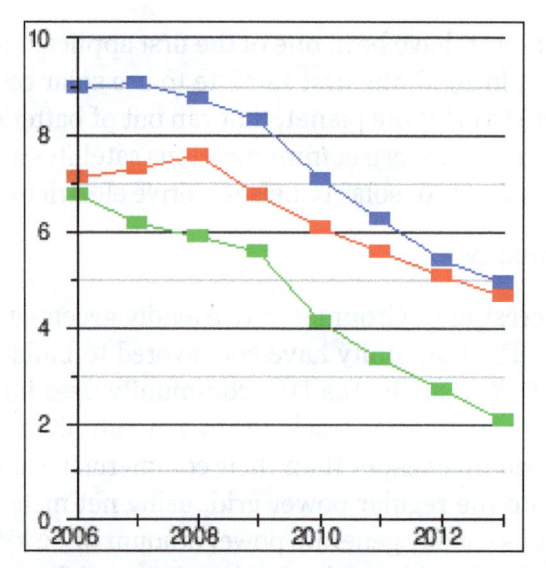

History of solar rooftop prices since 2006. Comparison in US$ per installed watt.

■ **Japan**   ■ **United States**   ■ **Germany**

The cost of producing photovotaic cells have dropped due to economies of scale in production and technological advances in manufacturing. For large-scale installations, prices below $1.00 per watt are now common. A price decrease of 50% had been achieved in Europe from 2006 to 2011 and there is a potential to lower the generation cost by 50% by 2020. Crystal silicon solar cells have largely been replaced by less expensive multicrystalline silicon solar cells, and thin film silicon solar cells have also been developed recently at lower costs of production. Although they are reduced in energy conversion efficiency from single crystalline "siwafers", they are also much easier to produce at comparably lower costs.

The table below shows the total cost in US cents per kWh of electricity generated by a photovoltaic system. The row headings on the left show the total cost, per peak kilowatt (kW_p), of a photovoltaic installation. Photovoltaic system costs have been declining and in Germany, for example, were reported to have fallen to USD 1389/kW_p by the end of 2014. The column headings across the top refer to the annual energy output in kWh expected from each installed kW_p. This varies by geographic region because the average insolation depends on the average cloudiness and the thickness of atmosphere traversed by the sunlight. It also depends on the path of the sun relative to the panel and the horizon. Panels are usually mounted at an angle based on latitude, and often they are adjusted seasonally to meet the changing solar declination. Solar tracking can also be utilized to access even more perpendicular sunlight, thereby raising the total energy output.

The calculated values in the table reflect the total cost in cents per kWh produced. They assume a 10% total capital cost (for instance 4% interest rate, 1% operating and maintenance cost, and depreciation of the capital outlay over 20 years). Normally, photovoltaic modules have a 25-year warranty.

Solar Power and Photovoltaics

Installation cost in $ per watt	Cost of generated kilowatt-hour by a PV-System (US¢/kWh) depending on solar radiation and installation cost during 20 years of operation									
	Insolation annually generated kilowatt-hours per installed kW-capacity (kWh/kW-p·y)								
	2400	2200	2000	1800	1600	1400	1200	1000	800
$0.20	0.8	0.9	1.0	1.1	1.3	1.4	1.7	2.0	2.5
$0.60	2.5	2.7	3.0	3.3	3.8	4.3	5.0	6.0	7.5
$1.00	4.2	4.5	5.0	5.6	6.3	7.1	8.3	10.0	12.5
$1.40	5.8	6.4	7.0	7.8	8.8	10.0	11.7	14.0	17.5
$1.80	7.5	8.2	9.0	10.0	11.3	12.9	15.0	18.0	22.5
$2.20	9.2	10.0	11.0	12.2	13.8	15.7	18.3	22.0	27.5
$2.60	10.8	11.8	13.0	14.4	16.3	18.6	21.7	26.0	32.5
$3.00	12.5	13.6	15.0	16.7	18.8	21.4	25.0	30.0	37.5
$3.40	14.2	15.5	17.0	18.9	21.3	24.3	28.3	34.0	42.5
$3.80	15.8	17.3	19.0	21.1	23.8	27.1	31.7	38.0	47.5
$4.20	17.5	19.1	21.0	23.3	26.3	30.0	35.0	42.0	52.5
$4.60	19.2	20.9	23.0	25.6	28.8	32.9	38.3	46.0	57.5
$5.00	20.8	22.7	25.0	27.8	31.3	35.7	41.7	50.0	62.5

USA	Japan	Germany	Small rooftop system cost and avg. insolation applied to data table, in 2013

Notes:
1. Cost per watt for rooftop system in 2013: Japan $4.64, United States $4.92, and Germany $2.05
2. Generated kilowatt-hour per installed watt-peak, based on average insolation for Japan (1500 kWh/m²/year), United States (5.0 to 5.5 kWh/m²/day), and Germany (1000 to 1200 kWh/m²/year).
3. A 2013 study by the Fraunhofer ISE concludes LCOE cost for a small PV system to be $0.16 (€0.12) rather than $0.22 per kilowatt-hour as shown in table (Germany).

System Cost 2013

In its 2014 edition of the "Technology Roadmap: Solar Photovoltaic Energy" report, the International Energy Agency (IEA) published prices in US$ per watt for residential, commercial and utility-scale PV systems for eight major markets in 2013.

Typical PV system prices in 2013 in selected countries (USD)								
USD/W	Australia	China	France	Germany	Italy	Japan	United Kingdom	United States
Residential	1.8	1.5	4.1	2.4	2.8	4.2	2.8	4.9
Commercial	1.7	1.4	2.7	1.8	1.9	3.6	2.4	4.5
Utility-scale	2.0	1.4	2.2	1.4	1.5	2.9	1.9	3.3

Source: IEA – Technology Roadmap: Solar Photovoltaic Energy report.

Regulation

Standardization

Increasing use of photovoltaic systems and integration of photovoltaic power into existing structures and techniques of supply and distribution increases the value of general standards and definitions for photovoltaic components and systems. The standards are compiled at the International Electrotechnical Commission (IEC) and apply to efficiency, durability and safety of cells, modules, simulation programs, plug connectors and cables, mounting systems, overall efficiency of inverters etc.

Planning and Permit

While article 690 of the National Electric Code provides general guidelines for the installation of photovoltaic systems, these guidelines may be superseded by local laws and regulations. Often a permit is required necessitating plan submissions and structural calculations before work may begin. Additionally, many locales require the work to be performed under the guidance of a licensed electrician. Check with the local City/County AHJ (Authority Having Jurisdiction) to ensure compliance with any applicable laws or regulations.

In the United States, the Authority Having Jurisdiction (AHJ) will review designs and issue permits, before construction can lawfully begin. Electrical installation practices must comply with standards set forth within the National Electrical Code (NEC) and be inspected by the AHJ to ensure compliance with building code, electrical code, and fire safety code. Jurisdictions may require that equipment has been tested, certified, listed, and labeled by at least one of the Nationally Recognized Testing Laboratories (NRTL). Despite the complicated installation process, a recent list of solar contractors shows a majority of installation companies were founded since 2000.

National Regulations

United Kingdom:

In the UK, PV installations are generally considered permitted development and don't require planning permission. If the property is listed or in a designated area (National Park, Area of Outstanding Natural Beauty, Site of Special Scientific Interest or Norfolk Broads) then planning permission is required.

United States:

In the US, many localities require a permit to install a photovoltaic system. A grid-tied system normally requires a licensed electrician to make the connection between the system and the grid-connected wiring of the building. Installers who meet these qualifications are located in almost every state. The State of California prohibits homeowners' associations from restricting solar devices.

Spain:

Although Spain generates around 40% of its electricity via photovoltaic and other renewable energy sources, and cities such as Huelva and Seville boast nearly 3,000 hours of sunshine per year, Spain has issued a solar tax to account for the debt created by the investment done by the Spanish government. Those who do not connect to the grid can face up to a fine of 30 million euros ($40 million USD).

Concentrated Solar Power

Concentrated solar power (also called concentrating solar power, concentrated solar thermal, and CSP) systems generate solar power by using mirrors or lenses to concentrate a large area of sunlight, or solar thermal energy, onto a small area. Electricity is generated when the concentrated light is converted to heat, which drives a heat engine (usually a steam turbine) connected to an electrical power generator or powers a thermochemical reaction (experimental as of 2013).

The three towers of the Ivanpah Solar Power Facility.

CSP is being widely commercialized and the CSP market saw about 740 megawatt (MW) of generating capacity added between 2007 and the end of 2010. More than half of this (about 478 MW) was installed during 2010, bringing the global total to 1095 MW. Spain added 400 MW in 2010, taking the global lead with a total of 632 MW, while the US ended the year with 509 MW after adding 78 MW, including two fossil–CSP hybrid plants. The Middle East is also ramping up their plans to install CSP based projects and as a part of that Plan, Shams-I which was the largest CSP Project in the world has been installed in Abu Dhabi, by Masdar. The largest CSP project in the world until January 2016 is Noor in Morocco and global operational power stands at 4,705 MW.

There is considerable academic and commercial interest internationally in a new form of CSP, called STEM, for off-grid applications to produce 24-hour industrial scale power for mining sites and remote communities in Italy, other parts of Europe, Australia, Asia, North Africa and Latin America. STEM uses fluidized silica sand as a thermal storage and heat transfer medium for CSP systems. It has been developed by Salerno-based Magaldi Industries. The first commercial application of STEM was scheduled to take place in Sicily from 2015.

CSP growth is expected to continue at a fast pace. As of January 2014, Spain had a total capacity of 2,300 MW making this country the world leader in CSP. United States follows with 1,740 MW.

Interest is also notable in North Africa and the Middle East, as well as India and China. In Italy, a handful of companies are trying to get authorization for 14 plants, totalling 392 MW, despite a strong local and political opposition. The global market has been dominated by parabolic-trough plants, which account for 90% of CSP plants.

CSP is not to be confused with concentrator photovoltaics (CPV). In CPV, the concentrated sunlight is converted directly to electricity via the photovoltaic effect.

Ivanpah in California is running 69 percent below advertised power output, and one Spanish company, Abengoa, that commercialized CSP both in the US and abroad, is teetering on the brink of bankruptcy. CSP technologies currently cannot compete on price with photovoltaics (solar panels), which have experienced huge growth in recent years due to falling prices of the panels. Another drawback of Ivanpah is that it lacks thermal energy storage, one of the big advantages that CSP has over PV and most other renewables, which require either large scale energy storage systems like pumped hydro or fast acting natural gas power plants to be there as backup for those times when they are not producing much energy.

History

A legend has it that Archimedes used a "burning glass" to concentrate sunlight on the invading Roman fleet and repel them from Syracuse. In 1973 a Greek scientist, Dr. Ioannis Sakkas, curious about whether Archimedes could really have destroyed the Roman fleet in 212 BC, lined up nearly 60 Greek sailors, each holding an oblong mirror tipped to catch the sun's rays and direct them at a tar-covered plywood silhouette 160 feet away. The ship caught fire after a few minutes; however, historians continue to doubt the Archimedes story.

In 1866, Auguste Mouchout used a parabolic trough to produce steam for the first solar steam engine. The first patent for a solar collector was obtained by the Italian Alessandro Battaglia in Genoa, Italy, in 1886. Over the following years, inventors such as John Ericsson and Frank Shuman developed concentrating solar-powered devices for irrigation, refrigeration, and locomotion. In 1913 Shuman finished a 55 HP parabolic solar thermal energy station in Maadi, Egypt for irrigation. The first solar-power system using a mirror dish was built by Dr. R.H. Goddard, who was already well known for his research on liquid-fueled rockets and wrote an article in 1929 in which he asserted that all the previous obstacles had been addressed.

Professor Giovanni Francia (1911–1980) designed and built the first concentrated-solar plant, which entered into operation in Sant'Ilario, near Genoa, Italy in 1968. This plant had the architecture of today's concentrated-solar plants with a solar receiver in the center of a field of solar collectors. The plant was able to produce 1 MW with superheated steam at 100 bar and 500 °C. The 10 MW Solar One power tower was developed in Southern California in 1981, but the parabolic-trough technology of the nearby Solar Energy Generating Systems (SEGS), begun in 1984, was more workable. The 354 MW SEGS is still the largest solar power plant in the world, and will remain so until the 390 MW Ivanpah power tower project reaches full power.

Current Technology

CSP is used to produce electricity (sometimes called solar thermoelectricity, usually generated through steam). Concentrated-solar technology systems use mirrors or lenses with tracking systems to focus a large area of sunlight onto a small area. The concentrated light is then used as heat or as a heat source for a conventional power plant (solar thermoelectricity). The solar concentrators used in CSP systems can often also be used to provide industrial process heating or cooling, such as in solar air-conditioning.

Concentrating technologies exist in five common forms, namely parabolic trough, enclosed trough, dish Stirlings, concentrating linear Fresnel reflector, and solar power tower. Although simple, these solar concentrators are quite far from the theoretical maximum concentration. For example, the parabolic-trough concentration gives about ⅓ of the theoretical maximum for the design acceptance angle, that is, for the same overall tolerances for the system. Approaching the theoretical maximum may be achieved by using more elaborate concentrators based on nonimaging optics.

Different types of concentrators produce different peak temperatures and correspondingly varying thermodynamic efficiencies, due to differences in the way that they track the sun and focus light. New innovations in CSP technology are leading systems to become more and more cost-effective.

Parabolic Trough

A parabolic trough consists of a linear parabolic reflector that concentrates light onto a receiver positioned along the reflector's focal line. The receiver is a tube positioned directly above the middle of the parabolic mirror and filled with a working fluid. The reflector follows the sun during the daylight hours by tracking along a single axis. A working fluid (e.g. molten salt) is heated to 150–350 °C (300–660 °F) as it flows through the receiver and is then used as a heat source for a power generation system. Trough systems are the most developed CSP technology. The Solar Energy Generating Systems (SEGS) plants in California, the world's first commercial parabolic trough plants, Acciona's Nevada Solar One near Boulder City, Nevada, and Andasol, Europe's first commercial parabolic trough plant are representative, along with Plataforma Solar de Almería's SSPS-DCS test facilities in Spain.

Enclosed Trough

Enclosed trough systems are used to absorb heat rather than process heat. The design encapsulates the solar thermal system within a greenhouse-like glasshouse. The glasshouse creates a protected environment to withstand the elements that can negatively impact reliability and efficiency of the solar thermal system. Lightweight curved solar-reflecting mirrors are suspended from the ceiling of the glasshouse by wires. A single-axis tracking system positions the mirrors to retrieve the optimal amount of sunlight. The mirrors concentrate the sunlight and focus it on a network of stationary steel pipes, also suspended from the glasshouse structure. Water is carried throughout the length of the pipe, which is boiled to generate steam when intense solar radiation is applied. Sheltering the mirrors from the wind allows them to achieve higher temperature rates and prevents dust from building up on the mirrors.

GlassPoint Solar, the company that created the Enclosed Trough design, states its technology can produce heat for Enhanced Oil Recovery (EOR) for about $5 per million British thermal units in sunny regions, compared to between $10 and $12 for other conventional solar thermal technologies.

Solar Power Tower

The PS10 solar power plant in Andalucía, Spain, concentrates sunlight from a field of heliostats onto a central solar power tower.

A solar power tower consists of an array of dual-axis tracking reflectors (heliostats) that concentrate sunlight on a central receiver atop a tower; the receiver contains a fluid deposit, which can consist of sea water. The working fluid in the receiver is heated to 500–1000 °C (773–1,273 K (932–1,832 °F)) and then used as a heat source for a power generation or energy storage system. An advantage of the solar tower is the reflectors can be adjusted instead of the whole tower. Power-tower development is less advanced than trough systems, but they offer higher efficiency and better energy storage capability. The Solar Two in Daggett, California and the CESA-1 in Plataforma Solar de Almeria Almeria, Spain, are the most representative demonstration plants. The Planta Solar 10 (PS10) in Sanlucar la Mayor, Spain, is the first commercial utility-scale solar power tower in the world. The 377 MW Ivanpah Solar Power Facility, located in the Mojave Desert, is the largest CSP facility in the world, and uses three power towers. The National Solar Thermal Test Facility, NSTTF located in Albuquerque, NM, is an experimental solar thermal test facility with a heliostat field capable of producing 6 MW.

Fresnel Reflectors

Fresnel reflectors are made of many thin, flat mirror strips to concentrate sunlight onto tubes through which working fluid is pumped. Flat mirrors allow more reflective surface in the same amount of space as a parabolic reflector, thus capturing more of the available sunlight, and they are much cheaper than parabolic reflectors. Fresnel reflectors can be used in various size CSPs.

Dish Stirling

A dish Stirling or dish engine system consists of a stand-alone parabolic reflector that concentrates light onto a receiver positioned at the reflector's focal point. The reflector tracks the Sun along two axes. The working fluid in the receiver is heated to 250–700 °C (480–1,300 °F) and then used by a Stirling engine to generate power. Parabolic-dish systems provide high solar-to-electric efficiency (between 31% and 32%), and their modular nature provides scalability. The Stirling Energy

Systems (SES), United Sun Systems (USS) and Science Applications International Corporation (SAIC) dishes at UNLV, and Australian National University's Big Dish in Canberra, Australia are representative of this technology. A world record for solar to electric efficiency was set at 31.25% by SES dishes at the National Solar Thermal Test Facility (NSTTF) in New Mexico on January 31, 2008, a cold, bright day. According to its developer, Ripasso Energy, a Swedish firm, in 2015 its Dish Sterling system being tested in the Kalahari Desert in South Africa showed 34% efficiency. The SES installation in Maricopa, Phoenix was the largest Stirling Dish power installation in the world until it was sold to United Sun Systems. Subsequently, larger parts of the installation have been moved to China as part of the huge energy demand.

Solar Thermal Enhanced Oil Recovery

Heat from the sun can be used to provide steam used to make heavy oil less viscous and easier to pump. Solar power tower and parabolic troughs can be used to provide the steam which is used directly so no generators are required and no electricity is produced. Solar thermal enhanced oil recovery can extend the life of oilfields with very thick oil which would not otherwise be economical to pump.

Deployment Around the World

Country	Total	Added
Spain	2,300	0
United States	1,634	+752
India	225	+175
United Arab Emirates	100	0
Algeria	25	0
Egypt	20	0
Morocco	20	0
Australia	12	0
China	10	0
Thailand	5	0
Source: REN21 Global Status Report, September 2015		

The commercial deployment of CSP plants started by 1984 in the US with the SEGS plants. The last SEGS plant was completed in 1990. From 1991 to 2005 no CSP plants were built anywhere in the world. Global installed CSP-capacity has increased nearly tenfold since 2004 and grew at an average of 50 percent per year during the last five years. In 2013, worldwide installed capacity increased by 36 percent or nearly 0.9 gigawatt (GW) to more than 3.4 GW. Spain and the United States remained the global leaders, while the number of countries with installed CSP were growing. There is a notable trend towards developing countries and regions with high solar radiation.

CSP is also increasingly competing with the cheaper photovoltaic solar power and with concentrator photovoltaics (CPV), a fast-growing technology that just like CSP is suited best for regions of high solar insolation. In addition, a novel solar CPV/CSP hybrid system has been proposed recently.

Worldwide Concentrated Solar Power (MW$_p$)														
Year	1984	1985	1989	1990	...	2006	2007	2008	2009	2010	2011	2012	2013	2014
Installed	14	60	200	80	0	1	74	55	179	307	629	803	872	925
Cumulative	14	74	274	354	354	355	429	484	663	969	1,598	2,553	3,425	4,400

Sources: REN21 · CSP-world.com · IRENA

Efficiency

The conversion efficiency η of the incident solar radiation into mechanical work – without considering the ultimate conversion step into electricity by a power generator – depends on the thermal radiation properties of the solar receiver and on the heat engine (e.g. steam turbine). Solar irradiation is first converted into heat by the solar receiver with the efficiency $\eta_{Receiver}$ and subsequently the heat is converted into work by the heat engine with the efficiency η_{Carnot}, using Carnot's principle. For a solar receiver providing a heat source at temperature T_H and a heat sink at room temperature T^0, the overall conversion efficiency can be calculated as follows:

$$\eta = \eta_{Receiver} \cdot \eta_{Carnot}$$

$$\text{with} \quad \eta_{Carnot} = 1 - \frac{T^0}{T_H}$$

$$\text{and} \quad \eta_{Receiver} = \frac{Q_{absorbed} - Q_{lost}}{Q_{solar}}$$

where Q_{solar}, $Q_{absorbed}$, Q_{lost} are respectively the incoming solar flux and the fluxes absorbed and lost by the system solar receiver.

For a solar flux I (e.g. I = 1000 W/m²) concentrated C times with an efficiency η_{Optics} on the system solar receiver with a collecting area A and an absorptivity α:

$$Q_{solar} = \eta_{Optics} ICA,$$

$$Q_{absorbed} = \alpha Q_{solar},$$

For simplicity's sake, one can assume that the losses are only radiative ones (a fair assumption for high temperatures), thus for a reradiating area A and an emissivity ϵ applying the Stefan-Boltzmann law yields:

$$Q_{lost} = A\epsilon\sigma T_H^4 C$$

Simplifying these equations by considering perfect optics $\eta_{Optics} = \alpha = 1$, then substituting in the first equation gives.

$$\eta = \left(1 - \frac{\sigma T_H^4}{IC}\right) \cdot \left(1 - \frac{T^0}{T_H}\right)$$

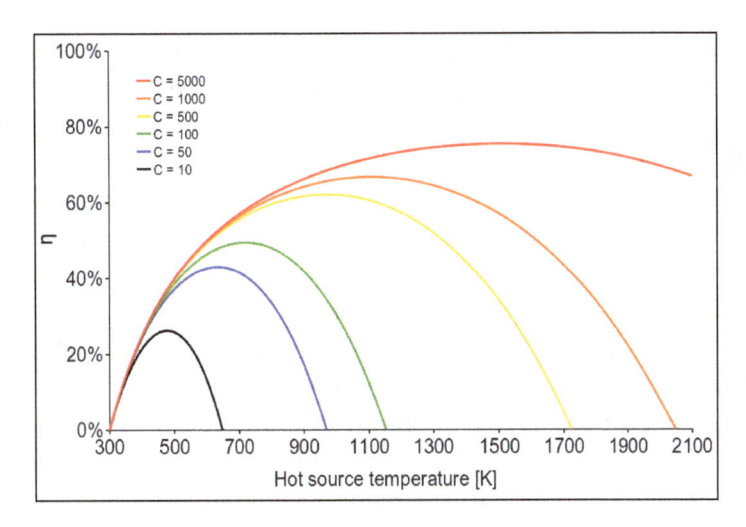

The graph shows that the overall efficiency does not increase steadily with the receiver's temperature. Although the heat engine's efficiency (Carnot) increases with higher temperature, the receiver's efficiency does not. On the contrary, the receiver's efficiency is decreasing, as the amount of energy it cannot absorb (Q_{lost}) grows by the fourth power as a function of temperature. Hence, there is a maximum reachable temperature. When the receiver efficiency is null (blue curve on the figure below), T_{max} is:

$$T_{max} = \left(\frac{IC}{\sigma}\right)^{0.25}$$

There is a temperature T_{opt} for which the efficiency is maximum, i.e. when the efficiency derivative relative to the receiver temperature is null:

$$\frac{d\eta}{dT_H}(T_{opt}) = 0$$

Consequently, this leads us to the following equation:

$$T_{opt}^5 - (0.75T^0)T_{opt}^4 - \frac{T^0 IC}{4\sigma} = 0$$

Solving this equation numerically allows us to obtain the optimum process temperature according to the solar concentration ratio C (red curve on the figure below).

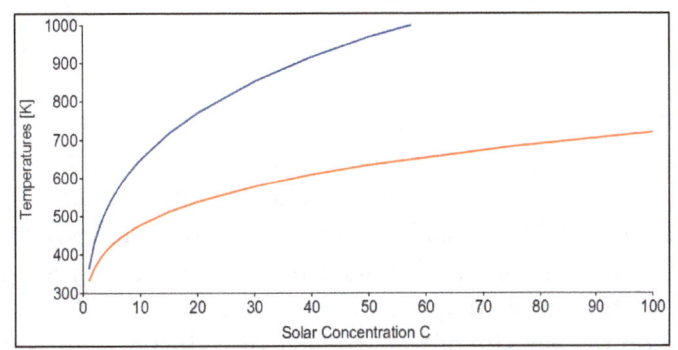

C	500	1000	5000	10000	45000 (max. for Earth)
T_{max}	1720	2050	3060	3640	5300
T_{opt}	970	1100	1500	1720	2310

Theoretical efficiencies aside, real-world experience of CSP reveals a 25-60 percent shortfall in projected production. A pilot 5 megawatt CSP power tower, Solar One, was converted to a 10 megawatt CSP power tower, Solar Two, decommissioned in 1999. Due to the success of Solar Two, a commercial power plant, called Solar Tres Power Tower, was built in Spain, renamed Gemasolar Thermosolar Plant. Gemasolar brilliant results have paved the way for the Crescent Dunes project. Ivanpah difficulties arise also from not having considered the lessons about the benefits of thermal storage. Solana in Arizona is 25 percent below projected numbers, Ivanpah in California, is 40 percent below projected numbers. A slightly bigger photovoltaic power station, like the 290 MW Agua Caliente Solar Project peaked at most to 741 GW·h in 2014, comparing with the 280 MW Solana growing 719 GW·h. Lower insolation related to a solar minimum is part of the problem, but there is clearly a technical issue with rating these systems or simply a commercial overstatement. Indeed, an other operator, that of the 280 MW Genesis Solar, projected only 580 GW·h production and instead made 621 GW·h in 2015. CSP once thought to be economically superior to photovoltaics, in 2015 it has proven not to be the case. Recent PV commercial power is selling for 1/3 recent CSP contracts.

Costs

As of 9 September 2009, the cost of building a CSP station was typically about US$2.50 to $4 per watt, while the fuel (the sun's radiation) is free. Thus a 250 MW CSP station would have cost $600–1000 million to build. That works out to $0.12 to 0.18 USD/kWh. New CSP stations may be economically competitive with fossil fuels. Nathaniel Bullard, a solar analyst at Bloomberg New Energy Finance, has calculated that the cost of electricity at the Ivanpah Solar Power Facility, a project under construction in Southern California, will be lower than that from photovoltaic power and about the same as that from natural gas. However, in November 2011, Google announced that they would not invest further in CSP projects due to the rapid price decline of photovoltaics. Google invested US$168 million on BrightSource. IRENA has published on June 2012 a series of studies titled: "Renewable Energy Cost Analysis". The CSP study shows the cost of both building and operation of CSP plants. Costs are expected to decrease, but there are insufficient installations to clearly establish the learning curve. As of March 2012, there were 1.9 GW of CSP installed, with 1.8 GW of that being parabolic trough.

Incentives

Spain

Solar-thermal electricity generation is eligible for feed-in tariff payments (art. 2 RD 661/2007), if the system capacity does not exceed the following limits: Systems registered in the register of systems prior to 29 September 2008: 500 MW for solar-thermal systems. Systems registered after 29 September 2008 (PV only). The capacity limits for the different system types are re-defined during the review of the application conditions every quarter (art. 5 RD 1578/2008, Annex III RD 1578/2008). Prior to the end of an application period, the market caps specified for each system type are published on the website of the Ministry of Industry, Tourism and Trade (art. 5 RD 1578/2008).

Since 27 January 2012, Spain has halted acceptance of new projects for the feed-in-tariff. Projects currently accepted are not affected, except that a 6% tax on feed-in-tariffs has been adopted, effectively reducing the feed-in-tariff.

Australia

At the federal level, under the Large-scale Renewable Energy Target (LRET), in operation under the Renewable Energy Electricity Act 2000, large scale solar thermal electricity generation from accredited RET power stations may be entitled to create large-scale generation certificates (LGCs). These certificates can then be sold and transferred to liable entities (usually electricity retailers) to meet their obligations under this tradeable certificates scheme. However, as this legislation is technology neutral in its operation, it tends to favour more established RE technologies with a lower levelised cost of generation, such as large scale onshore wind, rather than solar thermal and CSP. At State level, renewable energy feed-in laws typically are capped by maximum generation capacity in kWp, and are open only to micro or medium scale generation and in a number of instances are only open to solar PV (photovoltaic) generation. This means that larger scale CSP projects would not be eligible for payment for feed-in incentives in many of the State and Territory jurisdictions.

Future

A study done by Greenpeace International, the European Solar Thermal Electricity Association, and the International Energy Agency's SolarPACES group investigated the potential and future of concentrated solar power. The study found that concentrated solar power could account for up to 25% of the world's energy needs by 2050. The increase in investment would be from 2 billion euros worldwide to 92.5 billion euros in that time period. Spain is the leader in concentrated solar power technology, with more than 50 government-approved projects in the works. Also, it exports its technology, further increasing the technology's stake in energy worldwide. Because the technology works best with areas of high insolation (solar radiation), experts predict the biggest growth in places like Africa, Mexico, and the southwest United States. It indicates that the thermal storage systems based in nitrates (calcium, potassium, sodium,...) will make the CSP plants more and more profitable. The study examined three different outcomes for this technology: no increases in CSP technology, investment continuing as it has been in Spain and the US, and finally the true potential of CSP without any barriers on its growth. The findings of the third part are shown in the table below:

Year	Annual Investment	Cumulative Capacity
2015	21 billion euros	420 megawatts
2050	174 billion euros	1,500,000 megawatts

Finally, the study acknowledged how technology for CSP was improving and how this would result in a drastic price decrease by 2050. It predicted a drop from the current range of €0.23–0.15/kwh to €0.14–0.10/kwh. Recently the EU has begun to look into developing a €400 billion ($774 billion) network of solar power plants based in the Sahara region using CSP technology known as Desertec, to create "a new carbon-free network linking Europe, the Middle East and North Africa". The plan is backed mainly by German industrialists and predicts production of 15% of Europe's power by 2050. Morocco is a major partner in Desertec and as it has barely 1% of the electricity consumption of the EU, it will produce more than enough energy for the entire country with a large energy surplus to deliver to Europe.

Algeria has the biggest area of desert, and private Algerian firm Cevital has signed up for Desertec. With its wide desert (the highest CSP potential in the Mediterranean and Middle East regions ~ about 170 TWh/year) and its strategic geographical location near Europe Algeria is one of the key countries to ensure the success of Desertec project. Moreover, with the abundant natural-gas reserve in the Algerian desert, this will strengthen the technical potential of Algeria in acquiring Solar-Gas Hybrid Power Plants for 24-hour electricity generation.

Other organizations expect CSP to cost $0.06(US)/kWh by 2015 due to efficiency improvements and mass production of equipment. That would make CSP as cheap as conventional power. Investors such as venture capitalist Vinod Khosla expect CSP to continuously reduce costs and actually be cheaper than coal power after 2015.

On 9 September 2009; 6 years ago (2009-09-09), Bill Weihl, Google.org's green-energy spokesperson said that the firm was conducting research on the heliostat mirrors and gas turbine technology, which he expects will drop the cost of solar thermal electric power to less than $0.05/kWh in 2 or 3 years.

In 2009, scientists at the National Renewable Energy Laboratory (NREL) and SkyFuel teamed to develop large curved sheets of metal that have the potential to be 30% less expensive than today's best collectors of concentrated solar power by replacing glass-based models with a silver polymer sheet that has the same performance as the heavy glass mirrors, but at much lower cost and weight. It also is much easier to deploy and install. The glossy film uses several layers of polymers, with an inner layer of pure silver.

Telescope designer Roger Angel (Univ. of Arizona) has turned his attention to CPV, and is a partner in a company called Rehnu. Angel utilizes a spherical concentrating lens with large-telescope technologies, but much cheaper materials and mechanisms, to create efficient systems.

Recent experience with CSP technology in 2014 - 2015 at Solana in Arizona, and Ivanpah in Nevada indicate large production shortfalls in electricity generation between 25 and 40 percent. Producers blame clouds and stormy weather, but critics seem to think there are technological issues. These problems are causing utilities to pay inflated prices for wholesale electricity, and threaten the long-term viability of the technology. As photovoltaic costs continue to plummet, many think CSP has a limited future in utility-scale electricity production. On the other side, part of photovoltaic costs drop is the result of the fossil fuel cost reductions, which still powers most of PV production. Instead CSP would be unaffected when fossil prices return to higher quotes.

Very Large Scale Solar Power Plants

There are several proposals for gigawatt size, very large scale solar power plants. They include the Euro-Mediterranean Desertec proposal, Project Helios in Greece (10 gigawatt), and Ordos (2 gigawatt) in China. A 2003 study concluded that the world could generate 2,357,840 TWh each year from very large scale solar power plants using 1% of each of the world's deserts. Total consumption worldwide was 15,223 TWh/year (in 2003). The gigawatt size projects are arrays of single plants. The largest single plant in operation is the 370 MW Ivanpah Solar. In 2012, the BLM made available 97,921,069 acres (39,627,251 hectares) of land in the southwestern United States for solar projects, enough for between 10,000 and 20,000 gigawatts (GW).

Effect on Wildlife

It has been noted that insects can be attracted to the bright light caused by concentrated solar technology, and as a result birds that hunt them can be killed (burned) if the birds fly near the point where light is being focused onto. This can also affect raptors who hunt the birds. Federal wildlife officials have begun calling these power towers "mega traps" for wildlife.

However, the story about the Ivanpah Solar Power Facility was exaggerated, numbering the deaths in many tens of thousands, spreading alarm about concentrated solar power (CSP) plants, which was not grounded in facts, but on one opponent's speculation. According to rigorous reporting, in over six months, actually only 133 singed birds were counted. By focusing no more than 4 mirrors on any one place in the air during standby, at Crescent Dunes Solar Energy Project, in 3 months, the death rate dropped to zero fatalities.

Concentrator Photovoltaics

Concentrator photovoltaics (CPV) is a photovoltaic technology that generates electricity from sunlight. Contrary to conventional photovoltaic systems, it uses lenses and curved mirrors to focus sunlight onto small, but highly efficient, multi-junction (MJ) solar cells. In addition, CPV systems often use solar trackers and sometimes a cooling system to further increase their efficiency. Ongoing research and development is rapidly improving their competitiveness in the utility-scale segment and in areas of high insolation. This sort of solar technology can be thus used in smaller areas.

This Amonix system consists of thousands of small lenses, each focusing sunlight to ~500X higher intensity onto a tiny, high-efficiency multi-junction solar cell. A Tesla Roadster is parked beneath for scale.

Concentrator photovoltaics (CPV) modules on dual axis solar trackers in Golmud, China

Systems using high concentrator photovoltaics (HCPV) especially have the potential to become competitive in the near future. They possess the highest efficiency of all existing PV technologies, and a smaller photovoltaic array also reduces the balance of system costs. Currently, CPV is not used in the PV rooftop segment and is far less common than conventional PV systems. For regions with a high annual direct normal irradiance of 2000 kilowatt-hour (kWh) per square meter or more, the levelized cost of electricity is in the range of $0.08–$0.15 per kWh and installation cost for a 10-megawatt CPV power plant was identified to lie between €1.40–€2.20 (~$1.50-$2.30) per watt-peak (W_p).

In 2013 CPV installations accounted for only 0.1%, or 50 megawatts (MW), of the annual global PV market of nearly 39,000 MW. However, by the end of 2014, cumulative installations already amounted to 330 MW. Commercial HCPV systems reached efficiencies of up to 42% with concentration levels above 400, and the International Energy Agency sees potential to increase the efficiency of this technology to 50% by the mid-2020s. As of December 2014, the best lab cell efficiency for concentrator MJ-cells reached 46% (four or more junctions). Most CPV installations are located in China, the United States, South Africa, Italy and Spain.

HCPV directly competes with concentrated solar power (CSP) as both technologies are suited best for areas with high direct normal irradiance, which are also known as the Sun Belt region in the United States and the Golden Banana in Southern Europe. CPV and CSP are often confused with one another, despite being intrinsically different technologies from the start: CPV uses the photovoltaic effect to directly generate electricity from sunlight, while CSP – often called *concentrated solar thermal* – uses the heat from the sun's radiation in order to make steam to drive a turbine, that then produces electricity using a generator. Currently, CSP is more common than CPV.

History

Research into concentrator photovoltaics has taken place since the 1970s. Sandia National Laboratories in Albuquerque, New Mexico was the site for most of the early work, with the first modern photovoltaic concentrating system produced there late in the decade. Their first system was a linear-trough concentrator system that used a point focus acrylic Fresnel lens focusing on water-cooled silicon cells and two axis tracking. Ramón Areces' system, also developed in the late 1970s, used hybrid silicone-glass Fresnel lenses, while cooling of silicon cells was achieved with a passive heat sink.

Challenges

CPV systems operate most efficiently in concentrated sunlight, as long as the solar cell is kept cool through use of heat sinks. Diffuse light, which occurs in cloudy and overcast conditions, cannot be concentrated. To reach their maximum efficiency, CPV systems must be located in areas that receive plentiful direct sunlight.

The design of photovoltaic concentrators introduces a very specific optical design problem, with features that makes it different from any other optical design. It has to be efficient, suitable for mass production, capable of high concentration, insensitive to manufacturing and mounting inaccuracies, and capable of providing uniform illumination of the cell. All these reasons make nonimaging optics the most suitable for CPV.

CPV Strengths	CPV Weaknesses
High efficiencies for direct-normal irradiance	HCPV cannot utilize diffuse radiation. LCPV can only utilize a fraction of diffuse radiation
Low temperature coefficients	Tracking with sufficient accuracy and reliability is required
No cooling water required for passively cooled systems	May require frequent cleaning to mitigate soiling losses, depending on the site
Additional use of waste heat possible for systems with active cooling possible (e.g. large mirror systems)	Limited market – can only be used in regions with high DNI, cannot be easily installed on rooftops
Modular – kW to GW scale	Strong cost decrease of competing technologies for electricity production
Increased and stable energy production throughout the day due to tracking	Bankability and perception issues
Very low energy payback time	New generation technologies, without a history of production (thus increased risk)
Potential double use of land e.g. for agriculture, low environmental impact	Optical losses
High potential for cost reduction	Lack of technology standardization
Opportunities for local manufacturing	–
Smaller cell sizes could prevent large fluctuations in module price due to variations in semiconductor prices	–
Greater potential for efficiency increase in the future compared to single-junction flat plate systems could lead to greater improvements in land area use, BOS costs, and BOP costs	–
Source: Current Status of CPV report, January 2015. Table 2: Analysis of the strengths and weaknesses of CPV	

Efficiency

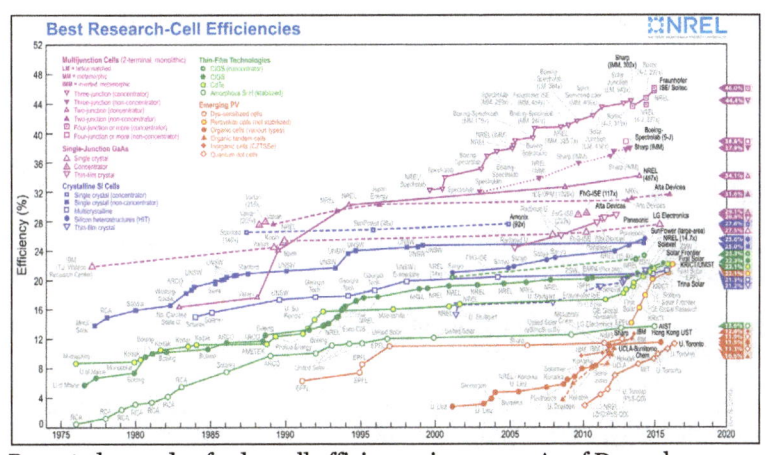

Reported records of solar cell efficiency since 1975. As of December 2014, best lab cell efficiency reached 46% (for ▫ multi-junction concentrator, 4+ junctions).

All CPV systems have a concentrating optic and a solar cell. Except for very low concentrations, active solar tracking is also necessary. Low concentration systems often have a simple booster reflector, which can increase solar electric output by over 30% from that of non-concentrator PV systems. Experimental results from such LCPV systems in Canada resulted in energy gains over 40% for prismatic glass and 45% for traditional crystalline silicon PV modules.

Semiconductor properties allow solar cells to operate more efficiently in concentrated light, as long as the cell Junction temperature is kept cool by suitable heat sinks. Efficiency of multi-junction photovoltaic cells developed in research is upward of 44% today, with the potential to approach 50% in the coming years.

Also crucial to the efficiency (and cost) of a CPV system is the concentrating optic since it collects and concentrates sunlight onto the solar cell. For a given concentration, nonimaging optics combine the widest possible acceptance angles with high efficiency and, therefore, are the most appropriate for use in solar concentration. For very low concentrations, the wide acceptance angles of nonimaging optics avoid the need for active solar tracking. For medium and high concentrations, a wide acceptance angle can be seen as a measure of how tolerant the optic is to imperfections in the whole system. It is vital to start with a wide acceptance angle since it must be able to accommodate tracking errors, movements of the system due to wind, imperfectly manufactured optics, imperfectly assembled components, finite stiffness of the supporting structure or its deformation due to aging, among other factors. All of these reduce the initial acceptance angle and, after they are all factored in, the system must still be able to capture the finite angular aperture of sunlight.

Grid Parity

Grid parity refers to the cost of solar/wind watt-hours produced compared to watt-hours available from the electrical utility grid. Grid parity is achieved when renewable energy watt-hours are monetarily equal to watt-hours produced on the grid (from coal, hydro, etc.).

Compared to conventional flat panel solar cells, CPV might be advantageous because the solar collector is less expensive than an equivalent area of solar cells. However CPV hardware (solar collector and tracker) is nearing US$1 per watt, whereas silicon flat panels that are commonly sold are now below $1 per watt (not including any associated power systems or installation charges).

References

- "Electric cars not solar panels, says Environment Commissioner". Parliamentary Commissioner for the Environment. 22 March 2016. Retrieved 22, June 2020
- Luque, Antonio & Hegedus, Steven (2003). Handbook of Photovoltaic Science and Engineering. John Wiley and Sons. ISBN 0-471-49196-9
- Worland, Justin (4 April 2016). "After years of torrid growth, residential solar power faces serious growing pains". Time. Vol. 187 no. 12. p. 24. Retrieved 17, April 2020
- "World's Largest CSP in Morocco Attracts South-South Learning | Climate Investment Funds". climateinvestmentfunds.org. Retrieved 13, February 2020

- "Solare Termodinamico, una ricchezza per il paese, p.21" (PDF). Associazione Nazionale Energia Solare Termodinamica (in Italian). Retrieved 27, July 2020
- Palz, Wolfgang (2013). Solar Power for the World: What You Wanted to Know about Photovoltaics. CRC Press. pp. 131–. ISBN 978-981-4411-87-5
- "Photovoltaic System Pricing Trends – Historical, Recent, and Near-Term Projections, 2014 Edition" (PDF). NREL. 22 September 2014. p. 4.

Chapter 5

Solar Energy Devices

The availability of modern solar appliances has resulted in the diverse applications of solar energy. From household devices to industrial sized machineries, solar energy can be used in diverse ways. Solar cookers, solar street lights, solar water heating systems, solar thermal pumps, solar refrigeration, solar distillation, solar dryer, etc. are some applications of solar energy which have been thoroughly discussed in this chapter.

Solar Cooker

A solar cooker consists of an:

- Insulated box of blackened aluminium in which utensils with food materials are kept.
- Reflector mirror hinged to one side of the box so that angle of reflector is adjusted.
- A glass layer consisting of two layers of clear window glass sheets.

The box is kept in such a way that solar radiation falls directly on glass cover. The reflector mirror is adjusted in that way additional solar radiation after mirror reflection is also incident on glass cover. The glass cover traps heat owing to green house effect providing more heating effect. The air temperature inside the box ranges from 140-160oC which is enough for boiling and heating purposes.

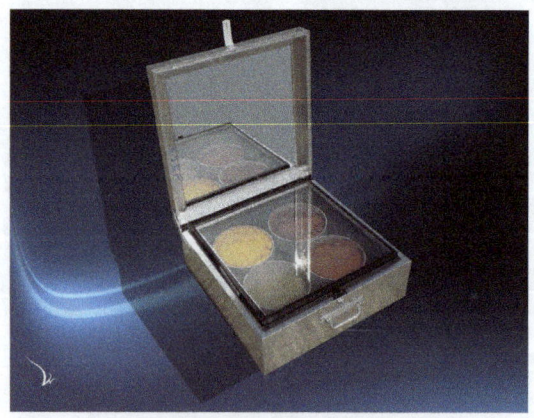

Solar cooker (Box type).

There are other types of solar cookers uses the solar collectors:

- Community solar cooker with arge reflectors concentrating solar radiation onto a black metal surface.
- Dish type solar cookers with parabolic reflectors.

Principles

1) Concentrating sunlight: A mirrored surface with high specular reflectivity is used to concentrate light from the sun on to a small cooking area. Depending on the geometry of the surface, sunlight can be concentrated by several orders of magnitude producing temperatures high enough to melt salt and smelt metal. For most household solar cooking applications, such high temperatures are not really required. Solar cooking products, thus, are typically designed to achieve temperatures of 150 °F (65 °C) (baking temperatures) to 750 °F (400 °C) (grilling/searing temperatures) on a sunny day.

2) Converting light energy to heat energy: Solar cookers concentrate sunlight onto a receiver such as a cooking pan. The interaction between the light energy and the receiver material converts light to heat. This conversion is maximized by using materials that conduct and retain heat. Pots and pans used on solar cookers should be matte black in color to maximize the absorption.

3) Trapping heat energy: It is important to reduce convection by isolating the air inside the cooker from the air outside the cooker. Simply using a glass lid on your pot enhances light absorption from the top of the pan and provides a greenhouse effect that improves heat retention and minimizes convection loss. This "glazing" transmits incoming visible sunlight but is opaque to escaping infrared thermal radiation. In resource constrained settings, a high-temperature plastic bag can serve a similar function, trapping air inside and making it possible to reach temperatures on cold and windy days similar to those possible on hot days.

Operation

Different kinds of solar cookers use somewhat different methods of cooking, but most follow the same basic principles.

Solar oven in use.

Food is prepared as if for an oven or stove top. However, because food cooks faster when it is in smaller pieces, food placed inside a solar cooker is usually cut into smaller pieces than it might otherwise be. For example, potatoes are usually cut into bite-sized pieces rather than roasted whole. For very simple cooking, such as melting butter or cheese, a lid may not be needed and the food may be placed on an uncovered tray or in a bowl. If several foods are to be cooked separately, then they are placed in different containers.

The container of food is placed inside the solar cooker, which may be elevated on a brick, rock, metal trivet, or other heat sink, and the solar cooker is placed in direct sunlight. Foods that cook quickly may be added to the solar cooker later. Rice for a mid-day meal might be started early in the morning, with vegetables, cheese, or soup added to the solar cooker in the middle of the morning. Depending on the size of the solar cooker and the number and quantity of cooked foods, a family may use one or more solar cookers.

A solar oven is turned towards the sun and left until the food is cooked. Unlike cooking on a stove or over a fire, which may require more than an hour of constant supervision, food in a solar oven is generally not stirred or turned over, both because it is unnecessary and because opening the solar oven allows the trapped heat to escape and thereby slows the cooking process. If wanted, the solar oven may be checked every one to two hours, to turn the oven to face the sun more precisely and to ensure that shadows from nearby buildings or plants have not blocked the sunlight. If the food is to be left untended for many hours during the day, then the solar oven is often turned to face the point where the sun will be when it is highest in the sky, instead of towards its current position.

The cooking time depends primarily on the equipment being used, the amount of sunlight at the time, and the quantity of food that needs to be cooked. Air temperature, wind, and latitude also affect performance. Food cooks faster in the two hours before and after the local solar noon than it does in either the early morning or the late afternoon. Large quantities of food, and food in large pieces, take longer to cook. As a result, only general figures can be given for cooking time. With a small solar panel cooker, it might be possible to melt butter in 15 minutes, to bake cookies in 2 hours, and to cook rice for four people in 4 hours. With a high performing parabolic solar cooker, you may be able to grill a steak in minutes. However, depending on local conditions and the solar cooker type, these projects could take half as long, or twice as long.

It is difficult to burn food in a solar cooker. Food that has been cooked even an hour longer than necessary is usually indistinguishable from minimally cooked food. The exception to this rule is some green vegetables, which quickly change from a perfectly cooked bright green to olive drab, while still retaining the desirable texture.

For most foods, such as rice, the typical person would be unable to tell how it was cooked from looking at the final product. There are some differences, however: Bread and cakes brown on their tops instead of on the bottom. Compared to cooking over a fire, the food does not have a smoky flavor.

Parabolic Reflectors

Parabolic solar cookers concentrate sunlight to a single point. When this point is focused on the bottom of a pot, it can heat the pot quickly to very high temperatures which can often be comparable with the temperatures achieved in gas and charcoal grills. These types of solar cookers are widely used in several regions of the world, most notably in China and India where hundreds of thousands of families currently use parabolic solar cookers for preparing food and heating water. Some parabolic solar cooker projects in China abate between 1-4 tons of carbon dioxide per year and receive carbon credits through the Clean Development Mechanism (CDM) and Gold Standard.

Solar tea kettle in Tibet.

Some parabolic solar cookers incorporate cutting edge materials and designs which lead to solar energy efficiencies >90%. Others are large enough to feed thousands of people each day, such as the solar bowl at Auroville in India, which makes 2 meals per day for 1,000 people.

If a reflector is axially symmetrical and shaped so its cross-section is a parabola, it has the property of bringing parallel rays of light (such as sunlight) to a point focus. If the axis of symmetry is aimed at the sun, any object that is located at the focus receives highly concentrated sunlight, and therefore becomes very hot. This is the basis for the use of this kind of reflector for solar cooking.

Paraboloidal Reflectors

A parabolic solar cooker with segmented construction.

Paraboloids are compound curves, which are more difficult to make with simple equipment than single curves. Although paraboloidal solar cookers can cook as well as or better than a conventional stove, they are difficult to construct by hand. Frequently, these reflectors are made using many small segments that are all single curves which together approximate compound curves.

Although paraboloids are difficult to make from flat sheets of solid material, they can be made quite simply by rotating open-topped containers which hold liquids. The top surface of a liquid which is being rotated at constant speed around a vertical axis naturally takes the form of a paraboloid. Centrifugal force causes material to move outward from the axis of rotation until a deep enough depression is formed in the surface for the force to be balanced by the levelling effect of gravity. It turns out that the depression is an exact paraboloid. If the material solidifies while it is rotating, the paraboloidal shape is maintained after the rotation stops, and can be used to make a reflector. This rotation technique is sometimes used to make paraboloidal mirrors for astronomical telescopes, and has also been used for solar cookers. Devices for constructing such paraboloids are known as rotating furnaces.

Paraboloidal reflectors generate high temperatures and cook quickly, but require frequent adjustment and supervision for safe operation. Several hundred thousand exist, mainly in China. They are especially useful for individual household and large-scale institutional cooking.

A Scheffler cooker. This reflector has an area of 16 m2 (170 sq ft), and concentrates 3 kW of heat.

A Scheffler cooker (named after its inventor, Wolfgang Scheffler) uses a large ideally paraboloidal reflector which is rotated around an axis that is parallel with the earth's using a mechanical mechanism, turning at 15 degrees per hour to compensate for the earth's rotation. The axis passes through the reflector's centre of mass, allowing the reflector to be turned easily. The cooking vessel is located at the focus which is on the axis of rotation, so the mirror concentrates sunlight onto it all day. The mirror has to be occasionally tilted about a perpendicular axis to compensate for the seasonal variation in the sun's declination. This perpendicular axis does not pass through the cooking vessel. Therefore, if the reflector were a rigid paraboloid, its focus would not remain stationary at the cooking vessel as the reflector tilts. To keep the focus stationary, the reflector's shape has to vary. It remains paraboloidal, but its focal length and other parameters change as it tilts. The Scheffler reflector is therefore flexible, and can be bent to adjust its shape. It is often made up of a large number of small plane sections, such as glass mirrors, joined together by flexible plastic. A framework that supports the reflector includes a mechanism that can be used to tilt it and also bend it appropriately. The mirror is never exactly paraboloidal, but it is always close enough for cooking purposes.

Sometimes the rotating reflector is located outdoors and the reflected sunlight passes through an opening in a wall into an indoor kitchen, often a large communal one, where the cooking is done.

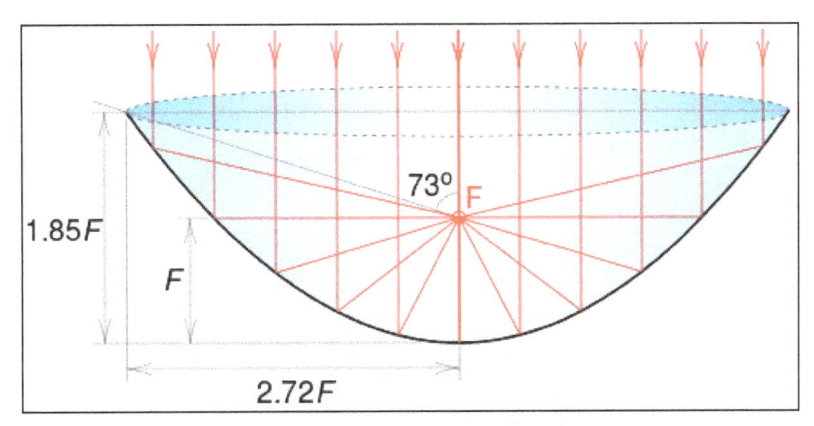

An oblique projection of a focus-balanced parabolic reflector.

Paraboloidal reflectors that have their centres of mass coincident with their focal points are useful. They can be easily turned to follow the sun's motions in the sky, rotating about any axis that passes through the focus. Two perpendicular axes can be used, intersecting at the focus, to allow the paraboloid to follow both the sun's daily motion and its seasonal one. The cooking pot stays stationary at the focus. If the paraboloidal reflector is axially symmetrical and is made of material of uniform thickness, its centre of mass coincides with its focus if the depth of the reflector, measured along its axis of symmetry from the vertex to the plane of the rim, is 1.8478 times its focal length. The radius of the rim of the reflector is 2.7187 times the focal length. The angular radius of the rim, as seen from the focal point, is 72.68 degrees.

Parabolic Troughs

Parabolic troughs are used to concentrate sunlight for solar-energy purposes. Some solar cookers have been built that use them in the same way. Generally, the trough is aligned with its focal line horizontal and east-west. The food to be cooked is arranged along this line. The trough is pointed so its axis of symmetry aims at the sun at noon. This requires the trough to be tilted up and down as the seasons progress. At the equinoxes, no movement of the trough is needed during the day to track the sun. At other times of year, there is a period of several hours around noon each day when no tracking is needed. Usually, the cooker is used only during this period, so no automatic sun tracking is incorporated into it. This simplicity makes the design attractive, compared with using a paraboloid. Also, being a single curve, the trough reflector is simpler to construct. However, it suffers from lower efficiency.

It is possible to use two parabolic troughs, curved in perpendicular directions, to bring sunlight to a point focus as does a paraboloidal reflector. The incoming light strikes one of the troughs, which sends it toward a line focus. The second trough intercepts the converging light and focuses it to a point.

Compared with a single paraboloid, using two partial troughs has important advantages. Each trough is a single curve, which can be made simply by bending a flat sheet of metal. Also, the light that reaches the targeted cooking pot is directed approximately downward, which reduces the danger of damage to the eyes of anyone nearby. On the other hand, there are disadvantages. More mirror material is needed, increasing the cost, and the light is reflected by two surfaces instead of one, which inevitably increases the amount that is lost.

The two troughs are held in a fixed orientation relative to each other by being both fixed to a frame. The whole assembly of frame and troughs has to be moved to track the sun as it moves in the sky. Commercially made cookers that use this method are available.In practical applications (like in car-headlights), concave mirrors are of parabolic shape.

Spherical Reflectors

The Solar Bowl in Auroville.

Spherical reflectors operate much like paraboloidal reflectors, such that the axis of symmetry is pointed towards the sun so that light is concentrated to a focus. However, the focus of a spherical reflector will not be a point focus because it suffers from a phenomenon known as spherical aberration. Some concentrating dishes (such as satellite dishes) that do not require a precise focus opt for a spherical curvature over a paraboloid. If the radius of the rim of spherical reflector is small compared with the radius of curvature of its surface (the radius of the sphere of which the reflector is a part), the reflector approximates a paraboloidal one with focal length equal to half of the radius of curvature.

Vacuum Tube Technology

Evacuated tube solar cookers are essentially a vacuum sealed between two layers of glass. The vacuum allows the tube to act both as a "super" greenhouse and an insulator. The central cooking tube is made from borosilicate glass, which is resistant to thermal shock, and has a vacuum beneath the surface to insulate the interior. The inside of the tube is lined with copper, stainless steel, and aluminum nitrile to better absorb and conduct heat from the sun's rays. Some vacuum tube solar cookers incorporate lightweight designs which allow great portability (such as the GoSun stove) Portable vacuum tube cookers such as the GoSun allow users to cook freshly caught fish on the beach without needing to light a fire.

Advantages

- High-performance parabolic solar cookers can attain temperatures above 290 °C (550 °F). They can be used to grill meats, stir-fry vegetables, make soup, bake bread, and boil water in minutes. Vacuum tube type cookers can heat up even in the clouds and freezing cold.

- Conventional solar box cookers attain temperatures up to 165 °C (325 °F). They can sterilize water or prepare most foods that can be made in a conventional oven or stove, including bread, vegetables and meat over a period of hours.

- Solar cookers use no fuel. This saves cost as well as reducing environmental damage caused by fuel use. Since 2.5 billion people cook on open fires using biomass fuels, solar cookers could have large economic and environmental benefits by reducing deforestation.

- When solar cookers are used outside, they do not contribute inside heat, potentially saving fuel costs for cooling as well. Any type of cooking may evaporate grease, oil, and other material into the air, hence there may be less cleanup.

Disadvantages

- Solar cookers are less useful in cloudy weather and near the poles (where the sun is low in the sky or below the horizon), so an alternative cooking source is still required in these conditions. Solar cooking advocates suggest three devices for an integrated cooking solution: a) a solar cooker; b) a fuel-efficient cookstove; c) an insulated storage container such as a basket filled with straw to store heated food. Very hot food may continue to cook for hours in a well-insulated container. With this three-part solution, fuel use is minimized while still providing hot meals at any hour, reliably.

- Some solar cookers, especially solar ovens, take longer to cook food than a conventional stove or oven. Using solar cookers may require food preparation start hours before the meal. However, it requires less hands-on time during the cooking, so this is often considered a reasonable trade-off.

- Cooks may need to learn special cooking techniques to fry common foods, such as fried eggs or flatbreads like chapatis and tortillas. It may not be possible to safely or completely cook some thick foods, such as large roasts, loaves of bread, or pots of soup, particularly in small panel cookers; the cook may need to divide these into smaller portions before cooking.

- Some solar cooker designs are affected by strong winds, which can slow the cooking process, cool the food due to convective losses, and disturb the reflector. It may be necessary to anchor the reflector, such as with string and weighted objects like bricks.

Solar Street Light

Solar street lights are effective and efficient light sources in which power is fed with the help of Photo-voltaic Panels.

They are generally mounted on the lighting structure.

There is a Rechargeable battery, which is charged by photo voltaic panels.

Then the charge of that battery is used to powers a fluorescent or Led Lamp during the night.

There have sensors, through them solar panels turn on and turn off automatically by sensing outdoor light with the help of light source.

They are designed to work at night.

Solar Street Light.

Working Principle

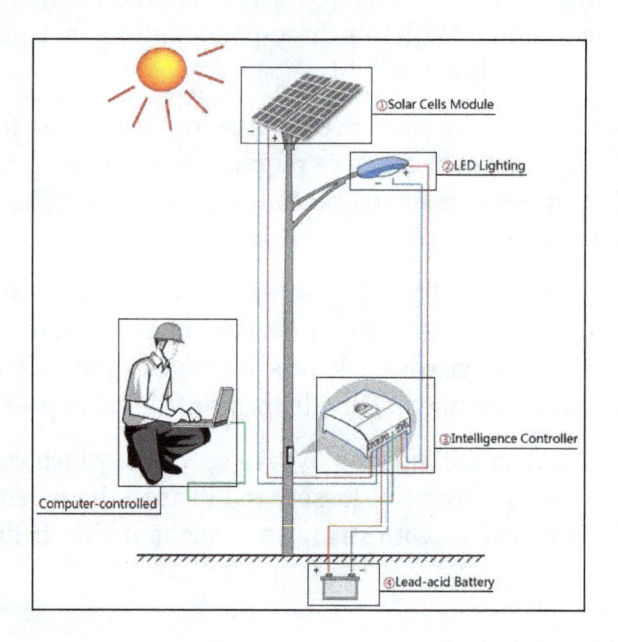

The Working Principle of Solar Street Light is very simple. Photo voltaic solar cells convert the radiation of sun light into electrical energy.

This conversion takes place by the use of the semiconductor material of the device.

This process of energy conversion is generally called the "Photo voltaic effect".

It is also known as solar cells, or "photo voltaic cells."

With the help of photo voltaic solar cells made of the principle effect of solar panels during the day. The received electrical energy stored in batteries.

At night when the illumination reduced to 10lux.

Then Solar cells board open the circuit voltage of about 4.5V.

Then charge and discharge controller is used to detect movement of the voltage value.

Charge and discharge controllers are generally used to protect the battery.

Components of Solar Street Lights

Solar Panel

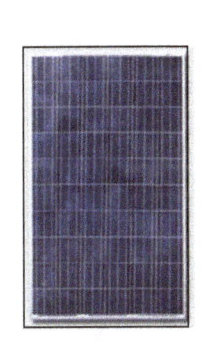

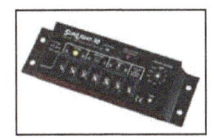

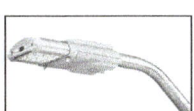

PV Panel.   Charge Controller.   Light Pole.   Photocell.   Battery.   LED Luminaire.

It is very important part of solar street lights.

Their main work is to convert solar energy into electricity.

There are 2 types of solar panel exists : Mono-crystalline and poly-crystalline.

The Conversion rate of mono-crystalline solar panel is much higher than poly-crystalline.

Lighting Fixture

Latest solar street light used LED as lighting source, because it provides much higher Lumens with lower consumption of power.

The energy consumption rate of LED fixture is at least 50% lower than HPS fixture.

Rechargeable Battery

The Rechargeable Battery stores the electricity from solar panel during the day and provides power to the fixture during night.

There are generally 2 types of batteries used:

a) Gel Cell Deep Cycle Battery and b) Lead Acid Battery.

Controller

Controller is also very important part for solar street light.

A controller is that circuit which decides when to switch on /off charging and lighting.

Pole

Strong Poles are required to solar street lights because there are very heavy components are mounted on the top of the pole like Fixtures, Panels and sometime batteries.

Advantages:

    The operation cost is not so high.

    This is a pollution free source of producing Electricity.

    NO external wires are used. so chances of occurrence of accidents are minimized.

    The parts of of this system are easy to carried. so remote access is applicable.

    They require less maintenance.

Disadvantages:

    The starting setup cost is high.

    Risk of theft is also higher.

    Rechargeable batteries are required to be changed many times.

    Not works in cloudy and rainy days.

    Snow and moisture effects its working.

Uses:

    They are the used for lightning at nights.

Solar Water Heating

A small capacity water heating system with natural circulation is as shown in figure. It is suitable to supply hot water for domestic purposes. It has two main component which include (i) flat plate collector to convert solar radiation in to heat energy and (ii) water storage tank to store hot water.

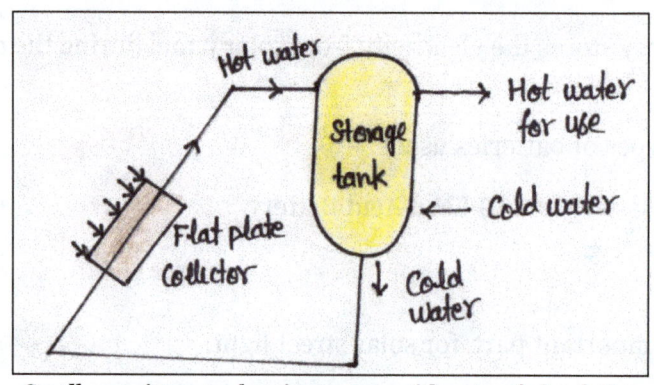

Small capacity water heating system with natural circulation.

The tank is located above the level of collector. The natural circulation of water is established from the collector to water tank and then from water tank to the collector. The hot water for use is withdrawn from the top of tank, which is replaced by cold water entering at the bottom of the tank. Water heating system is also provided with an auxiliary heating system so that the system can also work during cloudy and rainy days when sufficient solar radiation is unavailable.

Systems

Simple designs include a simple glass-topped insulated box with a flat solar absorber made of sheet metal, attached to copper heat exchanger pipes and dark-colored, or a set of metal tubes surrounded by an evacuated (near vacuum) glass cylinder. In industrial cases a parabolic mirror can concentrate sunlight on the tube. Heat is stored in a hot water storage tank. The volume of this tank needs to be larger with solar heating systems to compensate for bad weather and because the optimum final temperature for the solar collector is lower than a typical immersion or combustion heater. The heat transfer fluid (HTF) for the absorber may be water, but more commonly (at least in active systems) is a separate loop of fluid containing anti-freeze and a corrosion inhibitor delivers heat to the tank through a heat exchanger (commonly a coil of copper heat exchanger tubing within the tank). Copper is an important component in solar thermal heating and cooling systems because of its high heat conductivity, atmospheric and water corrosion resistance, sealing and joining by soldering and mechanical strength. Copper is used both in receivers and primary circuits (pipes and heat exchangers for water tanks).

Another lower-maintenance concept is the 'drain-back'. No anti-freeze is required; instead, all the piping is sloped to cause water to drain back to the tank. The tank is not pressurized and operates at atmospheric pressure. As soon as the pump shuts off, flow reverses and the pipes empty before freezing can occur.

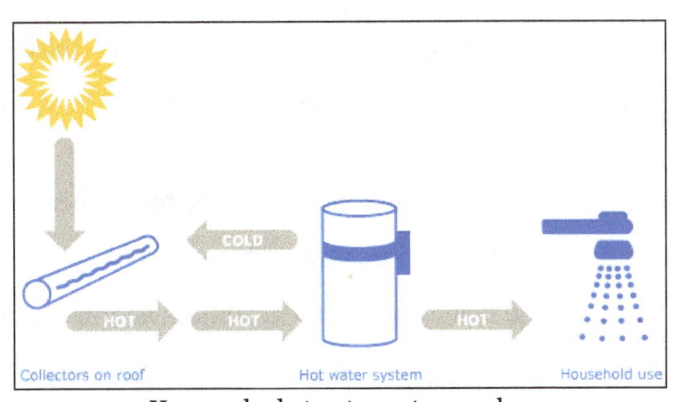

How a solar hot water system works.

Residential solar thermal installations fall into two groups: passive (sometimes called "compact") and active (sometimes called "pumped") systems. Both typically include an auxiliary energy source (electric heating element or connection to a gas or fuel oil central heating system) that is activated when the water in the tank falls below a minimum temperature setting, ensuring that hot water is always available. The combination of solar water heating and back-up heat from a wood stove chimney can enable a hot water system to work all year round in cooler climates, without the supplemental heat requirement of a solar water heating system being met with fossil fuels or electricity.

When a solar water heating and hot-water central heating system are used together, solar heat will either be concentrated in a pre-heating tank that feeds into the tank heated by the central heating, or the solar heat exchanger will replace the lower heating element and the upper element will remain to provide for supplemental heat. However, the primary need for central heating is at night and in winter when solar gain is lower. Therefore, solar water heating for washing and bathing is often a better application than central heating because supply and demand are better matched. In many climates, a solar hot water system can provide up to 85% of domestic hot water energy. This can include domestic non-electric concentrating solar thermal systems. In many northern European countries, combined hot water and space heating systems (solar combisystems) are used to provide 15 to 25% of home heating energy. When combined with storage, large scale solar heating can provide 50-97% of annual heat consumption for district heating.

Heat Transfer

Direct

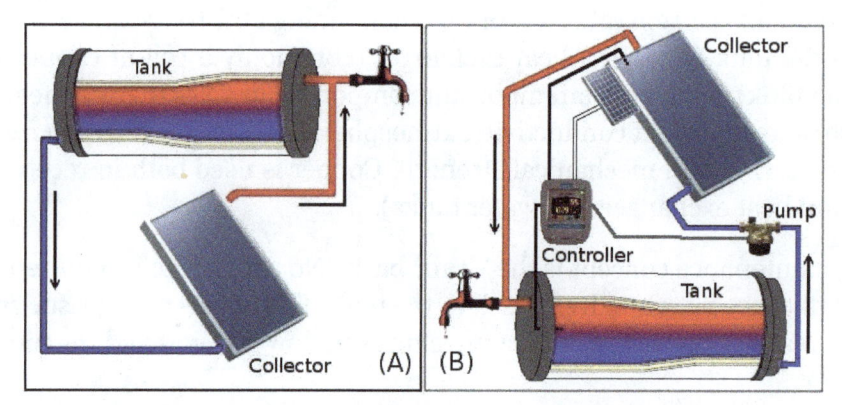

Direct systems: (A) Passive CHS system with tank above collector.
(B) Active system with pump and controller driven by a photovoltaic panel.

Direct or *open loop* systems circulate potable water through the collectors. They are relatively cheap. Drawbacks include:

- They offer little or no overheat protection unless they have a heat export pump.
- They offer little or no freeze protection, unless the collectors are freeze-tolerant.
- Collectors accumulate scale in hard water areas, unless an ion-exchange softener is used.

The advent of freeze-tolerant designs expanded the market for SWH to colder climates. In freezing conditions, earlier models were damaged when the water turned to ice, rupturing one or more components.

Indirect

Indirect or *closed loop* systems use a heat exchanger to transfer heat from the "heat-transfer fluid" (HTF) fluid to the potable water. The most common HTF is an antifreeze/water mix that typically uses non-toxic propylene glycol. After heating in the panels, the HTF travels to the heat exchanger, where its heat is transferred to the potable water. Indirect systems offer freeze protection and typically overheat protection.

Propulsion

Passive

Passive systems rely on heat-driven convection or heat pipes to circulate the working fluid. Passive systems cost less and require low or no maintenance, but are less efficient. Overheating and freezing are major concerns.

Active

Active systems use one or more pumps to circulate water and/or heating fluid. This permits a much wider range of system configurations.

Pumped systems are more expensive to purchase and to operate. However, they operate at higher efficiency can be more easily controlled.

Active systems have controllers with features such as interaction with a backup electric or gas-driven water heater, calculation and logging of the energy saved, safety functions, remote access and informative displays.

Passive Direct Systems

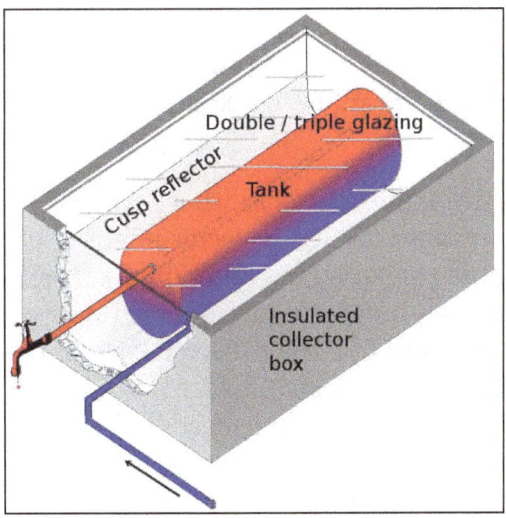

An integrated collector storage (ICS) system.

An *integrated collector storage* (ICS or batch heater) system uses a tank that acts as both storage and collector. Batch heaters are thin rectilinear tanks with a glass side facing the sun at noon. They are simple and less costly than plate and tube collectors, but they may require bracing if installed on a roof (to support 400–700 lb (180–320 kg) lbs of water), suffer from significant heat loss at night since the side facing the sun is largely uninsulated and are only suitable in moderate climates.

A *convection heat storage unit* (CHS) system is similar to an ICS system, except the storage tank and collector are physically separated and transfer between the two is driven by convection. CHS systems typically use standard flat-plate type or evacuated tube collectors. The storage tank must be located above the collectors for convection to work properly. The main benefit of CHS systems over ICS systems is that heat loss is largely avoided since the storage tank can be fully insulated.

Since the panels are located below the storage tank, heat loss does not cause convection, as the cold water stays at the lowest part of the system.

Active Indirect Systems

Pressurized antifreeze systems use a mix of antifreeze (almost always non-toxic propylene glycol) and water mix for HTF in order to prevent freeze damage.

Though effective at preventing freeze damage, antifreeze systems have drawbacks:

- If the HTF gets too hot the glycol degrades into acid and then provides no freeze protection and begins to dissolve the solar loop's components.
- Systems without drainback tanks must circulate the HTF – regardless of the temperature of the storage tank – to prevent the HTF from degrading. Excessive temperatures in the tank cause increased scale and sediment build-up, possible severe burns if a tempering valve is not installed, and if used for storage, possible thermostat failure.
- The glycol/water HTF must be replaced every 3–8 years, depending on the temperatures it has experienced.
- Some jurisdictions require more-expensive, double-walled heat exchangers even though propylene glycol is non-toxic.
- Even though the HTF contains glycol to prevent freezing, it circulates hot water from the storage tank into the collectors at low temperatures (e.g. below 40 °F (4 °C)), causing substantial heat loss.

A *drainback system* is an active indirect system where the HTF (usually pure water) circulates through the collector, driven by a pump. The collector piping is not pressurized and includes an open drainback reservoir that is contained in conditioned or semi-conditioned space. The HTF remains in the drainback reseervoir unless the pump is operating and returns there (emptying the collector) when the pump is switched off. The collector system, including piping, must drain via gravity into the drainback tank. Drainback systems are not subject to freezing or overheating. The pump operates only when appropriate for heat collection, but not to protect the HTF, increasing efficiency and reducing pumping costs.

Comparison

Characteristic	ICS (Batch)	Thermosi-phon	Active direct	Active indirect	Drainback	Bubble pump
Low profile-unobtrusive			✓	✓	✓	✓
Lightweight collector			✓	✓	✓	✓
Survives freezing weather			✓	✓	✓	✓
Low maintenance	✓	✓	✓		✓	✓

Simple: no ancillary control	✓	✓				✓
Retrofit potential to existing store			✓	✓	✓	✓
Space saving: no extra storage tank	✓	✓				
Comparison of SWH systems. Source: Solar Water Heating Basics—homepower.com'							

Components

Collector

Solar thermal collectors capture and retain heat from the sun and use it to heat a liquid. Two important physical principles govern the technology of solar thermal collectors:

- Any hot object ultimately returns to thermal equilibrium with its environment, due to heat loss from conduction, convection and radiation. Efficiency (the proportion of heat energy retained for a predefined time period) is directly related to heat loss from the collector surface. Convection and radiation are the most important sources of heat loss. Thermal insulation is used to slow heat loss from a hot object. This follows the Second law of thermodynamics (the 'equilibrium effect').

- Heat is lost more rapidly if the temperature difference between a hot object and its environment is larger. Heat loss is predominantly governed by the thermal gradient between the collector surface and the ambient temperatures. Conduction, convection and radiation all occur more rapidly over large thermal gradients (the delta-t effect).

Flat Plate

Flat plate collectors are an extension of the idea to place a collector in an 'oven'-like box with glass directly facing the Sun. Most flat plate collectors have two horizontal pipes at the top and bottom, called headers, and many smaller vertical pipes connecting them, called risers. The risers are welded (or similarly connected) to thin absorber fins. Heat-transfer fluid (water or water/antifreeze mix) is pumped from the hot water storage tank or heat exchanger into the collectors' bottom header, and it travels up the risers, collecting heat from the absorber fins, and then exits the collector out of the top header. Serpentine flat plate collectors differ slightly from this "harp" design, and instead use a single pipe that travels up and down the collector. However, since they cannot be properly drained of water, serpentine flat plate collectors cannot be used in drainback systems.

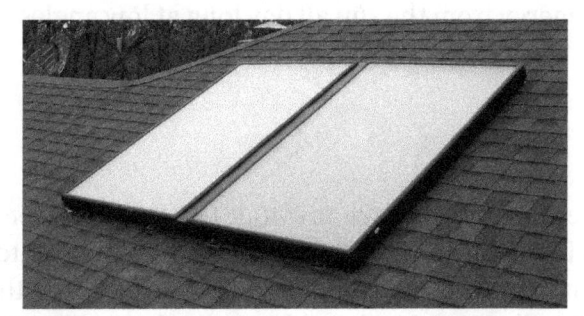

Flat-plate solar thermal collector, viewed from roof-level.

The type of glass used in flat plate collectors is almost always low-iron, tempered glass. Such glass can withstand significant hail without breaking, which is one of the reasons that flat-plate collectors are considered the most durable collector type.

Unglazed or formed collectors are similar to flat-plate collectors, except they are not thermally insulated nor physically protected by a glass panel. Consequently, these types of collectors are much less efficient. For pool heating applications, the water to be heated is often colder than the ambient roof temperature, at which point the lack of thermal insulation allows additional heat to be drawn from the surrounding environment.

Evacuated Tube

Evacuated tube solar water heater on a roof.

Evacuated tube collectors (ETC) are a way to reduce the heat loss, inherent in flat plates. Since heat loss due to convection cannot cross a vacuum, it forms an efficient isolation mechanism to keep heat inside the collector pipes. Since two flat glass sheets are generally not strong enough to withstand a vacuum, the vacuum is created between two concentric tubes. Typically, the water piping in an ETC is therefore surrounded by two concentric tubes of glass separated by a vacuum that admits heat from the sun (to heat the pipe) but that limits heat loss. The inner tube is coated with a thermal absorber. Vacuum life varies from collector to collector, from 5 years to 15 years.

Flat plate collectors are generally more efficient than ETC in full sunshine conditions. However, the energy output of flat plate collectors is reduced slightly more than ETCs in cloudy or extremely cold conditions. Most ETCs are made out of annealed glass, which is susceptible to hail, failing given roughly golf ball -sized particles. ETCs made from "coke glass," which has a green tint, are stronger and less likely to lose their vacuum, but efficiency is slightly reduced due to reduced transparency. ETCs can gather energy from the sun all day long at low angles due to their tubular shape.

Pump

PV Pump

One way to power an active system is via a photovoltaic (PV) panel. To ensure proper pump performance and longevity, the (DC) pump and PV panel must be suitably matched. Although a PV-powered pump does not operate at night, the controller must ensure that the pump does not operate when the sun is out but the collector water is not hot enough.

PV pumps offer the following advantages:

- Simpler/cheaper installation and maintenance.
- Excess PV output can be used for household electricity use or put back into the grid.
- Can dehumidify living space.
- Can operate during a power outage.
- Avoids the carbon consumption from using grid-powered pumps.

Bubble Pump

A bubble pump (also known as geyser pump) is suitable for flat panel as well as vacuum tube systems. In a bubble pump system, the closed HTF circuit is under reduced pressure, which causes the liquid to boil at low temperature as the sun heats it. The steam bubbles form a geyser, causing an upward flow. The bubbles are separated from the hot fluid and condensed at the highest point in the circuit, after which the fluid flows downward toward the heat exchanger caused by the difference in fluid levels. The HTF typically arrives at the heat exchanger at 70 °C and returns to the circulating pump at 50 °C. Pumping typically starts at about 50 °C and increases as the sun rises until equilibrium is reached.

Controller

A differential controller senses temperature differences between water leaving the solar collector and the water in the storage tank near the heat exchanger. The controller starts the pump when the water in the collector is sufficiently about 8–10 °C warmer than the water in the tank, and stops it when the temperature difference reaches 3–5 °C. This ensures that stored water always gains heat when the pump operates and prevents the pump from excessive cycling on and off. (In direct systems the pump can be triggered with a difference around 4 °C because they have no heat exchanger.)

Tank

The simplest collector is a water-filled metal tank in a sunny place. The sun heats the tank. This was how the first systems worked. This setup would be inefficient due to the equilibrium effect: as soon as heating of the tank and water begins, the heat gained is lost to the environment and this continues until the water in the tank reaches ambient temperature. The challenge is to limit the heat loss.

- The storage tank can be situated lower than the collectors, allowing increased freedom in system design and allowing pre-existing storage tanks to be used.
- The storage tank can be hidden from view.
- The storage tank can be placed in conditioned or semi-conditioned space, reducing heat loss.
- Drainback tanks can be used.

Insulated Tank

ICS or batch collectors reduce heat loss by thermally insulating the tank. This is achieved by encasing the tank in a glass-topped box that allows heat from the sun to reach the water tank. The other walls of the box are thermally insulated, reducing convection and radiation. The box can also have a reflective surface on the inside. This reflects heat lost from the tank back towards the tank. In a simple way one could consider an ICS solar water heater as a water tank that has been enclosed in a type of 'oven' that retains heat from the sun as well as heat of the water in the tank. Using a box does not eliminate heat loss from the tank to the environment, but it largely reduces this loss.

Standard ICS collectors have a characteristic that strongly limits the efficiency of the collector: a small surface-to-volume ratio. Since the amount of heat that a tank can absorb from the sun is largely dependent on the surface of the tank directly exposed to the sun, it follows that the surface size defines the degree to which the water can be heated by the sun. Cylindrical objects such as the tank in an ICS collector have an inherently small surface-to-volume ratio. Collectors attempt to increase this ratio for efficient warming of the water. Variations on this basic design include collectors that combine smaller water containers and evacuated glass tube technology, a type of ICS system known as an Evacuated Tube Batch (ETB) collector.

Applications

Evacuated Tube

ETSCs can be more useful than other solar collectors during winter season. ETCs can be used for heating and cooling purposes in industries like pharmaceutical and drug, paper, leather and textile and also for residential houses, hospitals nursing home, hotels swimming pool etc.

An ETC can operate at a range of temperatures from medium to high for solar hot water, swimming pool, air conditioning and solar cooker.

ETCs higher temperature (up to 200 °C (392 °F)) making them suitable for industrial applications such as steam generation, heat engine and solar drying.

Swimming Pools

Floating pool covering systems and separate STCs are used for pool heating.

Pool covering systems, whether solid sheets or floating disks, act as insulation and reduce heat loss. Much heat loss occurs through evaporation, and using a cover slows evaporation.

STCs for nonpotable pool water use are often made of plastic. Pool water is mildly corrosive due to chlorine. Water is circulated through the panels using the existing pool filter or supplemental pump. In mild environments, unglazed plastic collectors are more efficient as a direct system. In cold or windy environments evacuated tubes or flat plates in an indirect configuration are used in conjunction with a heat exchanger. This reduces corrosion. A fairly simple differential temperature controller is used to direct the water to the panels or heat exchanger either by turning a valve or operating the pump. Once the pool water has reached the required temperature, a diverter valve is used to return water directly to the pool without heating. Many systems are configured as drainback systems where the water drains into the pool when the water pump is switched off.

The collector panels are usually mounted on a nearby roof, or ground-mounted on a tilted rack. Due to the low temperature difference between the air and the water, the panels are often formed collectors or unglazed flat plate collectors. A simple rule-of-thumb for the required panel area needed is 50% of the pool's surface area. This is for areas where pools are used in the summer season only. Adding solar collectors to a conventional outdoor pool, in a cold climate, can typically extend the pool's comfortable usage by months and more if an insulating pool cover is used. An active solar energy system analysis program may be used to optimize the solar pool heating system before it is built.

Energy Footprint and Life Cycle Assessment

Energy Footprint

The source of electricity in an active SWH system determines the extent to which a system contributes to atmospheric carbon during operation. Active solar thermal systems that use mains electricity to pump the fluid through the panels are called 'low carbon solar'. In most systems the pumping reduces the energy savings by about 8% and the carbon savings of the solar by about 20%. However, low power pumps operate with 1-20W. Assuming a solar collector panel delivering 4 kWh/day and a pump running intermittently from mains electricity for a total of 6 hours during a 12-hour sunny day, the potentially negative effect of such a pump can be reduced to about 3% of the heat produced.

However, PV-powered active solar thermal systems typically use a 5–30 W PV panel and a small, low power diaphragm pump or centrifugal pump to circulate the water. This reduces the operational carbon and energy footprint.

Alternative non-electrical pumping systems may employ thermal expansion and phase changes of liquids and gases.

Life Cycle Energy Assessment

Recognised standards can be used to deliver robust and quantitative life cycle assessments (LCA). LCA considers the financial and environmental costs of acquisition of raw materials, manufacturing, transport, using, servicing and disposal of the equipment. Elements include:

- Financial costs and gains.
- Energy consumption.
- CO_2 and other emissions.

In terms of energy consumption, some 60% goes into the tank, with 30% towards the collector (thermosiphon flat plate in this case). In Italy, some 11 giga-joules of electricity are used in producing SWH equipment, with about 35% goes toward the tank, with another 35% towards the collector. The main energy-related impact is emissions. The energy used in manufacturing is recovered within the first 2–3 years of use (in southern Europe).

By contrast the energy payback time in the UK is reported as only 2 years. This figure was for a direct system, retrofitted to an existing water store, PV pumped, freeze tolerant and of 2.8 sqm aperture. For comparison, a PV installation took around 5 years to reach energy payback, according to the same comparative study.

In terms of CO_2 emissions, a large fraction of the emissions saved is dependent on the degree to which gas or electricity is used to supplement the sun. Using the Eco-indicator 99 points system as a yardstick (i.e. the yearly environmental load of an average European inhabitant) in Greece, a purely gas-driven system may have fewer emissions than a solar system. This calculation assumes that the solar system produces about half of the hot water requirements of a household.

A test system in Italy produced about 700 kg of CO2, considering all the components of manufacture, use and disposal. Maintenance was identified as an emissions-costly activity when the heat transfer fluid (glycol-based) was replaced. However, the emissions cost was recovered within about two years of use of the equipment.

In Australia, life cycle emissions were also recovered. The tested SWH system had about 20% of the impact of an electrical water heater and half that of a gas water heater.

Analysing their lower impact retrofit freeze-tolerant solar water heating system, Allen et al. (qv) reported a production CO2 impact of 337 kg, which is around half the environmental impact reported in the Ardente et al. (qv) study.

System Specification and Installation

- Most SWH installations require backup heating.
- The amount of hot water consumed each day must be replaced and heated. In a solar-only system, consuming a high fraction of the water in the reservoir implies significant reservoir temperature variations. The larger the reservoir the smaller the daily temperature variation.
- SWH systems offer significant scale economies in collector and tank costs. Thus the most economically efficient scale meets 100% of the heating needs of the application.
- Direct systems (and some indirect systems using heat exchangers) can be retrofitted to existing stores.
- Equipment components must be insulated to achieve full system benefits. The installation of efficient insulation significantly reduces heat loss.
- The most efficient PV pumps start slowly in low light levels, so they may cause a small amount of unwanted circulation while the collector is cold. The controller must prevent stored hot water from this cooling effect.
- Evacuated tube collector arrays can be adjusted by removing/adding tubes or their heat pipes, allowing customization during/after installation.
- Above 45 degrees latitude, roof mounted sun-facing collectors tend to outproduce wall-mounted collectors. However, arrays of wall-mounted steep collectors can sometimes produce more useful energy because gains in used energy in winter can offset the loss of unused (excess) energy in summer.

Solar Thermal Pump

Solar pumping utilizes the mechanical power generated by the solar radiation to run the water pump. Solar energy offers several beneficial features which make its utilization in irrigation pumping quite attractive. The features are as follows:

(i) The need for pumping arises most during the summer months when solar radiation is intense.

(ii) Pumping can be carried out intermittently without any problem.

(iii) Surplus pumped water can be stored in a reservoir or tank.

(iv) The requirement of water decreases during period of low radiation when solar pumping decrease. Evaporation losses reduce during cloudy days. Rainwater is also available during rainy days.

(v) There is relatively inexpensive running and maintenance cost.

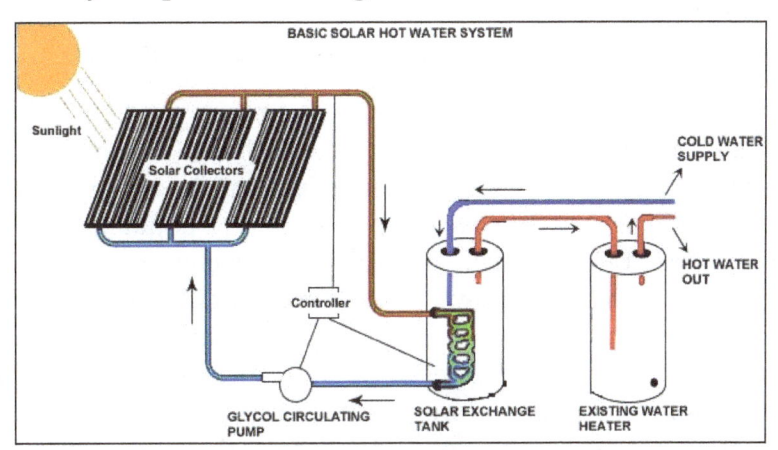

Solar Water Pump.

The solar pump is similar to solar heat engine working in low- temperature range. The source of heat engine works in low- temperature range. The source of heat is a solar collector. The heat is transported to a heat exchanger where heat is transferred to a refrigerant of low boiling point. The refrigerant evaporates and high- pressure vapour is taken to a turbine to do useful mechanical work by running the solar pump as shown in the figure. The outlet refrigerant vapour from turbine is condensed and takes to heat exchanger using feed for reuse.

Components

A photovoltaic solar powered pump system has three parts:

- Solar panels.
- The controller.
- The pump.

The solar panels make up most (up to 80%) of the systems cost. The size of the PV-system is directly dependent on the size of the pump, the amount of water that is required (m³/d) and the solar irradiance available.

The purpose of the controller is twofold. Firstly, it matches the output power that the pump receives with the input power available from the solar panels. Secondly, a controller usually provides a low voltage protection, whereby the system is switched off, if the voltage is too low or too high for the operating voltage range of the pump. This increases the lifetime of the pump thus reducing the need for maintenance.

Voltage of the solar pump motors can be AC (alternating current) or DC (direct current). Direct current motors are used for small to medium applications up to about 3 kW rating, and are suitable for applications such as garden fountains, landscaping, drinking water for livestock, or small irrigation projects. Since DC systems tend to have overall higher efficiency levels than AC pumps of a similar size, the costs are reduced as smaller solar panels can be used.

Finally, if an alternating current solar pump is used, an inverter is necessary that changes the direct current from the solar panels into alternating current for the pump. The supported power range of inverters extends from 0.15 to 55 kW and can be used for larger irrigation systems. However, the panel and inverters must be sized accordingly to accommodate the inrush characteristic of an AC motor.

Due to expensive prices of solar powered pumps, the market for solar pumps especially in developing countries such as India and Africa are highly dependent on government funding. "The Government of India has been taking strong steps towards solar pumps popularization since the past few years. The plan involves credit extensions to other developing countries to purchase solar pumps from manufacturers based in India. Moreover, the government also has plans to use the International Solar Alliance specifically for popularization of indigenously developed solar water pumps in Africa, with the help of NGOs. A few less developed countries in Africa are already experiencing the benefits of solar pumps installation in form of drastically increased farm yields and income of rural families. This will be the most important factor driving the market growth in India," opines an analyst at Future Market Insights.

Water Pumping

Solar powered water pumps can deliver drinking water as well as water for livestock or irrigation purposes. Solar water pumps may be especially useful in small scale or community based irrigation, as large scale irrigation requires large volumes of water that in turn require a large solar PV array. As the water may only be required during some parts of the year, a large PV array would provide excess energy that is not necessarily required, thus making the system inefficient.

Solar PV water pumping systems are used for irrigation and drinking water in India. The majority of the pumps are fitted with a 2000 watt - 3,700 watt motor that receives energy from a 4,800 Wp PV array. The 5hp systems can deliver about 124,000 liters of water/day from a total of 50 meters setoff head and 70 meters dynamic head. By 30 August 2016, a total of 1,20,000 solar PV water pumping systems have been installed in INDIA. in this system it produces 19M.H.W and 26 ton carbon dioxide.

Oil and Gas

In order to combat negative publicity related to the environmental impacts of fossil fuels, including fracking, the industry is embracing solar powered pumping systems. Many oil and gas wells require the accurate injection (metering) of various chemicals under pressure to sustain their operation and to improve extraction rates. Historically, these chemical injection pumps (CIP) have been driven by gas reciprocating motors utilizing the pressure of the well's gas and exhausting the raw gas into the atmosphere. Solar powered electrical pumps (solar CIP) can reduce these greenhouse gas emissions. Solar arrays (photovoltaic cells) not only provide a sustainable power source for the CIPs but can also provide an electric source to run remote SCADA type diagnostics with remote control and satellite/cell communications from very remote locations to a desktop or notebook monitoring computer.

Stirling Engine

Instead of generating electricity to turn a motor, sunlight can be concentrated on the heat exchanger of a Stirling engine and used to drive a pump mechanically. This dispenses with the cost of solar panels and electric equipment. In some cases the Stirling engine may be suitable for local fabrication, eliminating the difficulty of importing equipment. One form of Stirling engine is the fluidyne engine which operates directly on the pumped fluid as a piston. Fluidyne solar pumps have been studied since 1987 At least one manufacturer has conducted tests with a Stirling solar powered pump.

Solar Refrigeration

A simple solar operated absorption refrigeration system to cool a space is as shown in figure. The hot water transported from a flat plate collector is passed through a generator which is a heat exchanger.

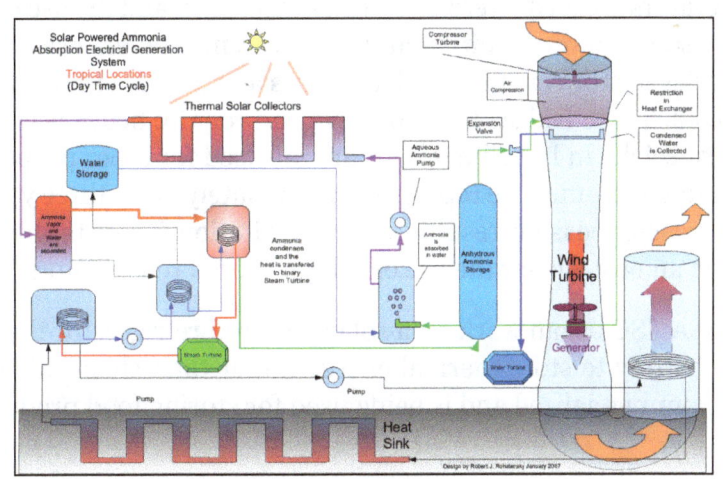

Solar Absorption refrigeration system.

The heat is transferred to a refrigerant and absorber solution. The refrigerant can be ammonia or water while absorber is water or lithium bromide which generates refrigerant vapours at high-pressure vapours are condensed into high- pressure liquid in the condenser. The high- pressure refrigerant liquid is throttled to low pressure and temperature by an expansion valve. The low pressure refrigerant takes heat from the evaporator and vapourises, thereby cooling air or water which can be used for cooling the space inside the building.

Naval Special Warfare support technicians receive training on a solar-powered refrigerator.

A solar-powered refrigerator is a refrigerator which runs on energy directly provided by sun, and may include photovoltaic or solar thermal energy.

Solar-powered refrigerators are able to keep perishable goods such as meat and dairy cool in hot climates, and are used to keep much needed vaccines at their appropriate temperature to avoid spoilage.

Solar-powered refrigerators are typically used in off-the-grid locations where utility provided AC power is not available.

Technology

Solar powered refrigerators are characterized by thick insulation and the use of a DC (not AC) compressor. Traditionally solar-powered refrigerators and vaccine coolers use a combination of solar panels and lead batteries to store energy for cloudy days and at night in the absence of sunlight to keep their contents cool. These fridges are expensive and require heavy lead-acid batteries which tend to deteriorate, especially in hot climates, or are misused for other purposes. In addition, the batteries require maintenance, must be replaced approximately every three years, and must be disposed of as hazardous wastes possibly resulting in lead pollution. These problems and the resulting higher costs have been an obstacle for the use of solar powered refrigerators in developing areas.

In the mid-1990s NASA JSC began work on a solar powered refrigerator that used phase change material rather than battery to store thermal energy rather than chemical energy. The resulting technology has been commercialized and is being used for storing food products and vaccines.

Use

Solar-powered refrigerators and other solar appliances are commonly used by individuals living off-the-grid. They provide a means for keeping food safe and preserved while avoiding a connection to utility-provided power. Solar refrigerators are also used in cottages and camps as an alternative to absorption refrigerators, as they can be safely left running year-round. Other uses include being used to keep medical supplies at proper temperatures in remote locations, and being used to temporarily store game at hunting camps.

Solar Distillation

It is the process to convert saline water into pure water using solar radiation. This is done with the help of the solar device known as solar still. A solar still consists of a shallow blackened basin filled with saline water that is to be distilled. It is covered with sloping transparent roof. The solar rays pass through transparent roof and are absorbed by the blackened surface of basin increasing temperature of the water. The water in the basin gets evaporated due to heat and rises to roof. The water drops or condensed water slip down along the sloping roof. This water is collected by the condensate channel and drained out from solar still.

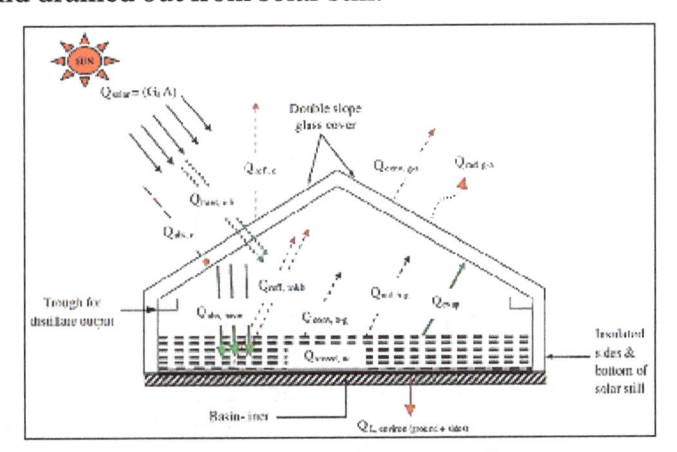

Working principle of Solar still.

Methods

In the direct method, a solar collector is coupled with a distilling mechanism and the process is carried out in one simple cycle. Solar stills of this type are described in survival guides, provided in marine survival kits, and employed in many small desalination and distillation plants. Water production by direct method solar distillation is proportional to the area of the solar surface and incidence angle and has an average estimated value of 3-4L/m2/day. Because of this proportionality and the relatively high cost of property and material for construction direct method distillation tends to favor plants with production capacities less than 200m³/day.

Indirect solar desalination employs two separate systems; a solar collection array, consisting of photovoltaic and/or fluid based thermal collectors, and a separate conventional desalination plant. Production by indirect method is dependent on the efficiency of the plant and the cost per unit produced is generally reduced by an increase in scale. Many different plant arrangements have been theoretically analyzed, experimentally tested and in some cases installed. They include but are not limited to multiple-effect humidification (MEH), multi-stage flash distillation (MSF), multiple-effect distillation (MED), multiple-effect boiling (MEB), humidification–dehumidification (HDH), reverse osmosis (RO), and freeze-effect distillation.

Indirect solar desalination systems using photovoltaic (PV) panels and reverse osmosis (RO) have been commercially available and in use since 2009. Output by 2013 is up to 1,600 litres (420 US gal) per hour per system, and 200 litres/day per square metre of PV panel. Municipal-scale systems are planned. Utirik Atoll in the Pacific Ocean has been supplied with fresh water this way since 2010.

Indirect solar desalination by a form of humidification/dehumidification is in use in the Seawater Greenhouse.

Types of Solar Desalination

There are two primary means of achieving desalination using solar energy, through a phase change by thermal input, or in a single phase through mechanical separation. Phase change (or multiphase) can be accomplished by either direct or indirect solar distillation. Single phase is predominantly accomplished by the use of photovoltaic cells to produce electricity to drive pumps although there are experimental methods being researched using solar thermal collection to provide this mechanical energy.

Multi-stage Flash Distillation (MSF)

Multi-stage flash distillation is one of the predominant conventional phase-change methods of achieving desalination. It accounts for roughly 45% of the total world desalination capacity and 93% of all thermal methods.

Solar derivatives have been studied and in some cases implemented in small and medium scale plants around the world. In Margarita de Savoya, Italy there is a 50–60 m^3/day MSF plant with a salinity gradient solar pond providing its thermal energy and storage capacity. In El Paso, Texas there is a similar project in operation that produces 19 m3/day. In Kuwait a MSF facility has been built using parabolic trough collectors to provide the necessary solar thermal energy to produce 100 m^3 of fresh water a day. And in Northern China there is an experimental, automatic, unmanned operation that uses 80 m^2 of vacuum tube solar collectors coupled with a 1 kW wind turbine (to drive several small pumps) to produce 0.8 m^3/day.

Production data shows that MSF solar distillation has an output capacity of 6-60 L/m^2/day versus the 3-4 L/m^2/day standard output of a solar still. MSF experience very poor efficiency during start up or low energy periods. In order to achieve the highest efficiency MSF requires carefully controlled pressure drops across each stage and a steady energy input. As a result, solar applications require some form of thermal energy storage to deal with cloud interference, varying solar patterns, night time operation, and seasonal changes in ambient air temperature. As thermal energy storage capacity increases a more continuous process can be achieved and production rates approach maximum efficiency.

Towered Desalination Plant Built in Pakistan

In 1993 a desalination plant was invented in Pakistan, producing 4 liters of water per square meter per day, which is at least ten times more productive than a conventional horizontal solar desalination plant. The structure is a raised tower made of concrete, with a tank at the top. The whole plant is covered with glass of the same shape, but slightly larger, allowing for a gap between the cement tower and the glass.

The tank is filled with saline water and water from an outside tank, drop by drop water enters the inner tank. The excessive water from the inner tank drips out onto the cement walls of the tower, from top to bottom. By solar radiation, the water on the wet surface and in the tank evaporate and condense on the inner surface of the glass cylinder and flow down onto the collecting drain channel. Meanwhile, the concentrated saline water drains out through a saline drain.

In this process fresh saline water is continuously added to the walls from the top of the tower. After evaporation, the remaining saline water falls down and drains out continuously. The movement of water also increases the energy of molecules and increases the evaporation process. The increase in the tower's height also increases the production.

Whereas in the conventional system water that is filled remains at a standstill for several days, a condenser is provided at the top in an isolated space, allowing cold water to pass through the condenser. The condensed hot vapors and hot water from the condenser are also thrown on the cement wall.

This plant's base is 3.5 by 1.5 by 10 feet (1.07 m × 0.46 m × 3.05 m) high, and gives about 12 litres (3.2 US gal) of water per day. Built horizontally, a structured plant receives solar radiation at noon only. But Zuberi's plant is a vertical tower and receives solar energy from sunrise till sunset. From early morning, it receives perpendicular radiation on one side of the plant, while at noon its top gets radiation equivalent to the horizontal plant. From noon till sunset, the other side receives maximum radiation.

By increasing the height, the tower plant receives more solar energy and the inner temperature increases as height increases. Ultimately this increases the water yield.

Different successive plants were constructed during the 1960s. A number of experiments have been conducted and a much more productive plant has been developed, with further work continuing.

This project can be implemented anywhere there is ground water, brine or sea water available with suitable sun. During different experiments a plant 6 feet (1.8 m) high can attain a temperature of 60 °C (140 °F), while a plant of 10 feet (3.0 m) high can reach a temperature of up to 86 °C (187 °F).

Solar Humidification–dehumidification

The solar humidification–dehumidification (HDH) process (also called the multiple-effect humidification–dehumidification process, solar multistage condensation evaporation cycle (SMCEC) or multiple-effect humidification (MEH), is a technique that mimics the natural water cycle on a shorter time frame by evaporating and condensing water to separate it from other substances. The driving force in this process is thermal solar energy to produce water vapor which is later condensed in a separate chamber. In sophisticated systems, waste heat is minimized by collecting the heat from the condensing water vapor and pre-heating the incoming water source. This system is effective for small- to mid- scale desalination systems in remote locations because of the relative inexpensiveness of solar thermal collectors.

Problems with Thermal Systems

There are two inherent design problems facing any thermal solar desalination project. Firstly, the system's efficiency is governed by preferably high heat and mass transfer rates during evaporation and condensation. The surfaces have to be properly designed within the contradictory objectives of heat transfer efficiency, economy, and reliability.

Secondly, the heat of condensation is valuable because it takes large amounts of solar energy to evaporate water and generate saturated, vapor-laden hot air. This energy is, by definition, transferred to the condenser's surface during condensation. With most forms of solar stills, this heat of condensation is ejected from the system as waste heat. The challenge still existing in the field today, is to achieve the optimum temperature difference between the solar-generated vapor and the seawater-cooled condenser, maximal reuse of the energy of condensation, and minimizing the asset investment.

Solutions for Thermal Systems

Efficient desalination systems use heat recovery to allow the same heat input to provide several times the water than a simple evaporative process such as solar stills.

One solution to the barrier presented by the high level of solar energy required in solar desalination efforts is to reduce the pressure within the reservoir. This can be accomplished using a vacuum pump, and significantly decreases the temperature of heat energy required for desalination. For example, water at a pressure of 0.1 atmospheres boils at 50 °C (122 °F) rather than 100 °C (212 °F). The atmospheric pressure creates a force of 9 metric tons/square meter (the difference between 1 atm outside and 0.1 atm inside) and so the total vertical force applied by the dome on the concrete wall (without any weight of the dome itself!) should be about 7.2 million tons on this wall, that means 2300 tons per meter of the concrete wall. More: it is impossible to build a dome able to resist.

Solar Dryer

This is used to dry the food items with the use of solar radiation. The drying can be achieved directly or by indirectly.

Direct

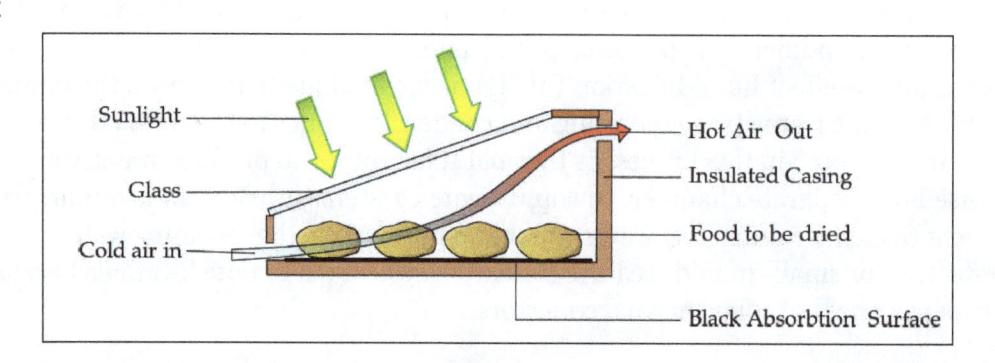

Direct solar dryers expose the substance to be dehydrated to direct sunlight. Historically, food and clothing was dried in the sun by using lines, or laying the items on rocks or on top of tents. In Mongolia cheese and meat are still traditionally dried using the top of the ger (tent) as a solar dryer. In these systems the solar drying is assisted by the movement of the air (wind) that removes the more saturated air away from the items being dried. More recently, complex drying racks and solar tents were constructed as solar dryers.

One modern type of solar dryer has a black absorbing surface which collects the light and converts it to heat; the substance to be dried is placed directly on this surface. These driers may have enclosures, glass covers and/or vents to in order to increase efficiency.

Indirect

Industrial indirect solar fruit and vegetable dryer.

In indirect solar dryers, the black surface heats incoming air, rather than directly heating the substance to be dried. This heated air is then passed over the substance to be dried and exits upwards often through a chimney, taking moisture released from the substance with it. They can be very simple, just a tilted cold frame with black cloth to an insulated brick building with active ventilation and a back-up heating system. One of the advantages of the indirect system is that it is easier to protect the food, or other substance, from contamination whether wind-blown or by birds, insects, or animals. Also, direct sun can chemically alter some foods making them less appetizing.

References

- Wethe, David (29 November 2012). "For Fracking, It›s Getting Easier Being Green». Bloomberg Businessweek. Retrieved 12, March 2020

- McDermott, James E. Horne; Maura (2001). The next green revolution : essential steps to a healthy, sustainable agriculture. New York [u.a.]: Food Products Press. p. 226. ISBN 1560228865

- Li, Wei; Rubin, Tzameret H.; Onyina, Paul A. (2013-05-01). "Comparing Solar Water Heater Popularization Policies in China, Israel and Australia: The Roles of Governments in Adopting Green Innovations". Sustainable Development. 21 (3): 160–170. doi:10.1002/sd.1547. ISSN 1099-1719

- "A Performance Calculator for Grid-Connected PV Systems». Rredc.nrel.gov. Archived from the original on January 18, 2012. Retrieved 09, January 2020

- Linda Frederick Yaffe (2007). Solar Cooking for Home and Camp. Mechanicsburg, PA: Stackpole Books. pp. 16–20. ISBN 0-8117-3402-1
- Fincher, Johnathan. "GoSun:Portable solar oven cooks food in as little as 10 minutes". Retrieved 23, July 2020
- H. M. Healey (2007). "Economics of Solar". Cogeneration & Distributed Generation Journal. 22 (3): 35–49. doi:10.1080/15453660709509122

Permissions

All chapters in this book are published with permission under the Creative Commons Attribution Share Alike License or equivalent. Every chapter published in this book has been scrutinized by our experts. Their significance has been extensively debated. The topics covered herein carry significant information for a comprehensive understanding. They may even be implemented as practical applications or may be referred to as a beginning point for further studies.

We would like to thank the editorial team for lending their expertise to make the book truly unique. They have played a crucial role in the development of this book. Without their invaluable contributions this book wouldn't have been possible. They have made vital efforts to compile up to date information on the varied aspects of this subject to make this book a valuable addition to the collection of many professionals and students.

This book was conceptualized with the vision of imparting up-to-date and integrated information in this field. To ensure the same, a matchless editorial board was set up. Every individual on the board went through rigorous rounds of assessment to prove their worth. After which they invested a large part of their time researching and compiling the most relevant data for our readers.

The editorial board has been involved in producing this book since its inception. They have spent rigorous hours researching and exploring the diverse topics which have resulted in the successful publishing of this book. They have passed on their knowledge of decades through this book. To expedite this challenging task, the publisher supported the team at every step. A small team of assistant editors was also appointed to further simplify the editing procedure and attain best results for the readers.

Apart from the editorial board, the designing team has also invested a significant amount of their time in understanding the subject and creating the most relevant covers. They scrutinized every image to scout for the most suitable representation of the subject and create an appropriate cover for the book.

The publishing team has been an ardent support to the editorial, designing and production team. Their endless efforts to recruit the best for this project, has resulted in the accomplishment of this book. They are a veteran in the field of academics and their pool of knowledge is as vast as their experience in printing. Their expertise and guidance has proved useful at every step. Their uncompromising quality standards have made this book an exceptional effort. Their encouragement from time to time has been an inspiration for everyone.

The publisher and the editorial board hope that this book will prove to be a valuable piece of knowledge for students, practitioners and scholars across the globe.

Index

A
Air Mass, 21-22
Alternating Current, 51, 129, 169, 173, 222
Amorphous Silicon, 148, 155, 157
Atmospheric Effects, 54

B
Beam Radiation, 21-22, 27-28, 35, 97, 122, 124-125

C
Chromosphere, 1-4
Concentrated Photovoltaics, 51, 131, 142, 155, 172, 178
Concentrated Solar Power, 76, 116, 129-131, 133, 141, 143, 155, 163, 177-178, 185, 190, 193-196
Concentrator Photovoltaics, 131, 142, 159, 176, 178, 186, 190, 195-196
Convective Zone, 2, 8, 100-101, 107
Crystalline Silicon, 49, 141, 149, 152, 155, 157, 164-165, 169, 198

D
Decay Product, 16
Declination Angle, 22-23, 25-26
Diffuse Radiation, 9, 21, 27-28, 197
Direct Current, 51, 129, 148, 173, 178-179, 222
Direct Normal Irradiance, 196
Diverter Valve, 218

E
Electric Motor, 45
Electrical Grid, 51, 143-144, 151, 164, 166, 174, 176, 179
Electromagnetic Radiation, 3, 9, 43, 50, 68
Evacuated Tube Collector, 67-69, 98, 220
Evaporation, 81, 83, 99-101, 108, 142, 160, 178, 218, 221, 227-228
Extraterrestrial Radiation, 15, 27

F
Flat Plate Collector, 67-68, 99, 102, 210, 223
Flat-plate Collector, 120
Floatovoltaic, 160, 178

G
Greenhouse Gas, 140, 223
Grid Parity, 137, 151, 155-156, 165, 198

H
Heat Energy, 4, 7, 12-13, 45, 50, 64-65, 67, 94, 96, 116, 131, 201, 210, 215, 228
Heat Loss, 70, 75, 98-99, 101, 117-119, 213-218, 220
Heat Sink, 155, 190, 196, 202
Heat Storage, 59, 64-65, 83, 89, 99-100, 117, 213
Heat Transfer Fluid, 70, 87-88, 94, 99, 116, 126, 211, 220
Heat Transfer Medium, 78, 80, 82, 85, 88-89, 116, 119, 185
Hour Angle, 23-26

I
Infrared Radiation, 43
Irradiance, 9-10, 13-14, 28-30, 32-40, 42, 70, 95, 112, 168, 170, 177, 196-197, 222
Irradiation, 13-14, 31, 82-83, 101, 153, 167, 190

L
Lead Acid, 179, 209
Light Energy, 47-48, 76, 116, 148, 201
Linear Fresnel Reflector, 69, 71, 75, 83, 85, 91, 93, 130, 187

M
Maximum Power Point Tracking, 52, 150, 174, 179
Molten Salt, 59, 80, 82-83, 85, 87, 89, 94, 116, 119, 124, 143-144, 175, 187
Monocrystalline Silicon, 148

O
Optical Efficiency, 96-98, 102, 104, 122

P
Parabolic Reflector, 72, 130, 187-188, 205
Parabolic Solar Cooker, 202-203
Parabolic Trough Collector, 69, 77, 80
Photodiode, 31, 34-35, 37-39, 148
Photovoltaic Array, 36, 148, 168, 170, 174, 196
Polycrystalline Silicon, 148
Pyranometer, 28-37, 39-40, 42, 176
Pyrheliometer, 28, 40-42

R
Radiant Energy, 13, 40, 99, 113, 127
Radiation Spectrum, 9, 13-14

Index

Radionuclides, 15-16

Radon, 15-20

Reflectors, 67, 72-73, 75-76, 91-93, 111, 116, 130, 188, 200, 202-206

Renewable Energy, 43, 46-47, 55, 111, 116, 129, 137-140, 143, 147-148, 151, 155, 163, 168, 175, 184, 192-194, 198

S

Solar Azimuth Angle, 24

Solar Boiler, 93

Solar Cell, 47-51, 111, 114-115, 129, 142, 150, 153, 155, 164, 169, 171, 195-198

Solar Cells, 47, 50-51, 111, 129-131, 141-142, 146-150, 152-153, 155-159, 162, 164, 168, 176, 181, 195, 198, 208-209

Solar Collector, 42, 67, 83, 90, 97, 99, 119, 121, 159, 186, 198, 211, 217, 219, 221, 225

Solar Concentrator, 29, 95

Solar Constant, 7, 9, 15, 54-55

Solar Cooker, 200-203, 207, 218

Solar Dryers, 229

Solar Pond, 99-109, 120, 226

Solar Power, 1, 47, 51, 54, 56, 76, 90-91, 93, 95, 111, 115-116, 129-137, 139-141, 143-144, 146, 150-151, 155-159, 161-165, 172, 175, 177-178, 185-190, 192-196, 198-199

Solar Radiation, 9, 12-15, 21, 27-28, 30-31, 34-35, 37, 40-45, 58, 67-70, 99-102, 109, 112, 131, 159, 176, 178, 183, 187, 189-190, 193, 200, 210-211, 221, 225, 227-228

Solar Spectrum, 10, 51, 77-78, 111-112, 150

Solar Tracker, 45-46, 92, 142, 168, 172

Solar Water Heater, 69-70, 218, 229

Solar Water Heating, 68-69, 200, 210-212, 215, 220

Stirling Engine, 95, 124, 130, 188, 223

Summer Solstice, 11

T

Thermal Energy Storage, 58-59, 70, 94, 134, 143, 186, 226

Thermal System, 187

Thermosyphon, 45, 72

W

Water Vapour, 15

Winter Solstice, 11

Z

Zenith Angle, 23-24, 35